Mordecai Cubitt Cooke

Romance of Low Life amongst Plants

Facts and Phenomena of Cryptogamic Vegetation

Mordecai Cubitt Cooke

Romance of Low Life amongst Plants
Facts and Phenomena of Cryptogamic Vegetation

ISBN/EAN: 9783744774482

Printed in Europe, USA, Canada, Australia, Japan

Cover: Foto ©berggeist007 / pixelio.de

More available books at **www.hansebooks.com**

ROMANCE OF LOW LIFE
AMONGST PLANTS.

FACTS AND PHENOMENA
OF CRYPTOGAMIC VEGETATION

BY

M. C. COOKE, M.A., LL.D., A.L.S.,

AUTHOR OF

"FREAKS AND FANCIES OF PLANT LIFE," "PONDS AND DITCHES,"
"VEGETABLE WASPS AND PLANT WORMS," ETC.

PUBLISHED UNDER THE DIRECTION OF THE GENERAL LITERATURE
COMMITTEE.

LONDON:
SOCIETY FOR PROMOTING CHRISTIAN KNOWLEDGE,
NORTHUMBERLAND AVENUE, W.C.;
43, QUEEN VICTORIA STREET, E.C.
BRIGHTON: 135, NORTH STREET.
NEW YORK: E. & J. B. YOUNG AND CO.
1893.

PREFACE.

THE growing interest in the lower forms of
vegetable life, which has been stimulated
during the past thirty or forty years by improvements
in the microscope and in methods of manipulation,
has led me to believe that an account of some of
the most remarkable of the facts and phenomena of
cryptogamic vegetation would prove of considerable
interest to young readers, and even suggestive to
older ones. It is not so much to the results of new or
recent investigations that I have desired to give pre-
dominant interest, as to the general influence which
increased knowledge of the structure and history of
these minute and obscure plants has had upon the
romantic beliefs and unsubstantial theories of a less
enlightened age. It may be true that many of the
facts now collected are not absolutely new, but may
be found scattered through learned treatises, or dispersed over scientific journals; yet in that condition
they were not available for general reading, nor were
they likely to come under the notice of any except
those specially interested in each particular branch of

botanical study. In order to present these facts in a concise form, divested of their more technical surroundings, I have collected them into a volume for general reading, in the hope of exciting some interest, where no interest was felt before, in the wonders of the unseen world. Some of them are "curiosities of vegetable life," some are exploded theories or wild romance, and some will come as revelations of the Master's work in the most minute of His creations; but all will furnish material for thought and reflection, and teach by fact—more convincing than argument. To present, in a popular manner, a few of the phenomena of plant life, as exhibited in the simplest and lowest forms, can scarce need further apology.

M. C. COOKE.

CONTENTS.

———⋇———

vi CONTENTS.

CONTENTS.

ROMANCE OF LOW LIFE
AMONGST PLANTS.

—⋅∗⋅—

CRYPTOGAMIC VEGETATION.

THE entire vegetable world of great and small, from the Australian gum tree to the smallest microbe, is divisible, for scientific purposes, into two primary sections, which are termed respectively the Phanerogamia, or flowering plants, and the Cryptogamia, or flowerless plants ; but it is only with the latter that we are called upon to deal. It may be premised that they are designated as *flowerless* plants, because the organs of fructification are not developed into conspicuous flowers, as in the Phanerogamia, and are even so elementary in some cases that they are reduced to a simple ovary, or ovum, or reproductive cell.

For general purposes, it is customary to speak of one section of Cryptogamic plants as the Higher Cryptogamia, inferring that they approach nearest to flowering plants ; and the other section as the Lower Cryptogamia, intimating that it embraces the lowest, or simplest forms. The former includes

B

the Ferns, and similar plants, the Mosses, and the Liverworts; the latter includes Lichens, Fungi, and Algæ, or water-weeds. The Higher Cryptogamia are possessed of an axis, more or less complete, descending into rootlets, or representatives of rootlets, and furnished with foliaceous expansions, the analogue of leaves. In the ferns the resemblance to leaves is nearly complete, but their function is more complex, since they bear the organs of fructification on the under side, or some portion of their surface. The Lower Cryptogamia are much more variable in form, without terminal vegetation or rootlets, often very minute, and may even be reduced to a single cell. This must suffice as a general introduction, since more explicit description would lead us away from the main object of this work.

It cannot be claimed for these humble vegetables that they emulate the glories of the primeval forest, wherein league after league of towering trees gives shelter and food to myriads of animals, and is eloquent with the song of birds. It cannot be urged on their behalf that they present the elegant forms, the splendid coloration, or the delicious odours of the innumerable flowers which festoon the trunks of the giants of the forest, or carpet the soil with a living arabesque. It cannot be written of them that they supply the staple food of the human race, the chief support of the animal world, and, either directly or indirectly, are the source of most of the essentials, and many of the luxuries, of civilized life. Nor can it be vaunted of them that they are the staples of a commerce which represents the wealth of the world, and extends from pole to pole. On the contrary they are sometimes despised for their insignificance, rejected for their lack of utility, and trampled to death for their ubiquitous presence as the scavengers and regenerators of the vegetable world.

But they are not wholly lacking in their contribution even to the beauty of vegetation. In all countries the ferns supply a green and graceful undergrowth, or garnish decaying trunks. The mosses cover, what would otherwise be naked and barren tracts, with a soft velvety carpet of a fresh and lively green, or disguise the ugliness of fallen and lifeless trunks by covering them with a new garment of life. The naked rocks, up to the limits of eternal snow, the bleak moorland, and crumbling ruins of a past generation are redeemed from dreariness by the perennial lichens. The waters of the ocean, capable of supporting no other forms of vegetable life, are redeemed from sterility by the presence everywhere of the sea-weeds, which furnish food and shelter to myriads of aquatic animals. Even the fungi, in most cases rather destroyers than decorators, contribute something, in their more persistent forms, to the variety, harmony, and even the beauty of woodland vegetation.

Apart from all this, it must not be forgotten that, even if they are to be relegated to a subsidiary position as earth's decorators, the Cryptogamia have a mission to fulfil, which may be as important, if not as evident, as that of the Phanerogamia. It is possible to demonstrate that the whole aquatic animal life is as dependent upon aquatic vegetation as terrestrial animal life is dependent, directly or indirectly, upon terrestrial vegetation.   It could be indicated in what manner fungi are essential to the disintegration of dead matter, and its rejuvenescence in new forms of vegetable life.   Past, present, and future, we may rest assured, that the humblest individuals of the lowest of the Cryptogamia have been, are, and ever will be, as essential in the great unity, which we call Nature, as the loftiest forest tree, or the most odoriferous of beautiful flowers.

> " These are Thy wonders, Lord of power,
> Killing and quickening, bringing down to hell
> And up to heaven in an hour ;
> Making a chiming of a passing bell.
> We say amiss,
> This or that is :
> Thy Word is all, if we could spell."

NUMBER FOUR.

It is well known that in Flowering Plants there is a typical number which represents the parts of the floral organs, and those of reproduction.   In the larger group of Dicotyledonous plants, those having two seed-leaves, this number is *five*, which is represented in the five petals, five sepals, five stamens (or some multiple of five), five carpels, etc.   In Monocotyledonous plants the number is *three*, which in like manner is represented in the floral organs and in the fruit.   There are, of course, some exceptions, but not sufficient to affect the general result.

In the Cryptogamia there is also a prevalent or typical number, in which the organs, or products, of reproduction are chiefly concerned, and, in this case the number is *four*.

Descending to particulars, we have in some of the ferns, as in *Aspidium filix mas*, each spore mother-cell provided with a nucleus, and, in consequence of its division, two new large clear nuclei arise, and between them a line of separation is sometimes apparent. After division of these nuclei, four new smaller nuclei appear, the mother-cell splitting up into four spore cells. In *Marsilea* the frond is four-lobed in all the species. In *Selaginella* the macrosporangia contain usually four (sometimes two or eight) macrospores. In mosses the number of spores produced is indefinite, but they arise in fours. The capsule is entire in true mosses, but in the *Andreaceæ* the capsule splits into four valves, as it does also in the *Hepaticæ*, or liver-worts. In the latter the mode of division of the mother-cells of the spores into four varies; anyway, cell walls are formed, and the mother-cell breaks up into four spores.

In the Lower Cryptogamia the Lichens are pro-vided with asci, which sometimes contain four sporidia, and sometimes eight, or other multiple of four. In Fungi the vast number of *Ascomycetes*, which have the sporidia contained in asci, are sometimes tetra-sporous, more often octosporous, and occasionally some other multiple of four, such as sixteen or thirty-two. The very large order of *Hymenomycetes*, in which the spores are naked, and borne at the apex of basidia, normally have four spores to each basidium. In the *Gastromycetes*, or Puff-ball Family, Mr. Massee ob-serves that the number of spores is not so constant as in the *Hymenomycetes*. In the *Phalloideæ* they are from four to eight; in *Bovista*, *Lycoperdon*, and

Tylostoma they are normally four. The Algæ, or
sea-weeds, are no exception, for in the *Floridcæ* the
sexual organs of reproduction are gonidia, four of
which are usually formed in a mother-cell, and hence
termed " tetragonidia." In some genera of the
Fucaceæ the protoplasm of the oögonium divides
into two, four, or eight. In the *Palmellaceæ* and the
Confervoid Algæ a quaternary division is so general,
both in vegetation and reproduction, that a glance
through any volume of analytical plates will furnish
a profusion of examples in support of the universality
of "number four," and prove it to be the typical
number for the Cryptogamia.

It has been suggested that this quaternary arrange-
ment is related to the method of growth by fissura-
tion, which prevails in the lower cryptogamia. A
simple cell divides in one direction into two, and
then in an opposite direction into four; each of these
latter units, in progressive growth, repeats the same
process, and so on *ad infinitum*. This does not seem
sufficient to account for the production of four or
eight sporidia in a membranous sac, or the four
valves in the capsules of *Andræa* or the *Hepaticæ*,
but it may have something to do with the division of
compound conidia, or sporidia, into two, four, six,
or eight cells. If we advert to the moulds, for
example, the globose or oval spores (in which the
length is scarcely equal to the breadth) are usually
continuous, but when the length is equal to, or
exceeds the diameter, there is a tendency towards
septation, which seems to increase with the length
of the spore. The elongated conidium, or spore,

becomes at first bicellular, by transverse division in the centre, then each cell is subdivided in like manner, and the spore becomes four-celled ; less frequently, by biseptation of each cell it is six-celled, or by partition of each one of four cells it becomes eight-celled. Very rarely, when the septation is complete does each cell exceed its own diameter in length, and as rarely is the number of cells three, five, or seven.

In the ascomycetous fungi, the sporidia, to the number of four, or eight—rarely two or sixteen—are contained in membranous sacs, but never three or five. These sporidia, in the majority of instances, are continuous, but in some at first continuous and afterwards two or more celled. In all cases, whether in the moulds or the ascomycetes, each cell possesses all the properties and functions of an individual spore ; so that, after all, the reproductive unit is reduced to a single cell. The same remarks apply also to the lichens, which are, at least, the analogues of the ascomycetous fungi.

In support of the hypothesis that compound spore, in fungi are families of individuals, as in algæ, we may quote *Phragmidium* amongst the uredines, and *Fusarium* amongst the moulds. In the former, each of the four to eight cells is perforate, and produces its own germ-tube, at the end of which the small protospores are developed, each cell being as much an individual as the single cell in *Uromyces*. In the latter there is sometimes even a dehiscence at the joints, whilst each cell germinates independently, and can reproduce the original mould as completely as

does the entire unicellular spore. Collaterally, it may
be added, that each cell of such a muriformly septate
spore as that of *Macrosporium* amongst moulds, and
Pleospora in the *Ascomycetes*, is, for all practical
purposes, a germinating spore, and that the so-called
spore is a family, or colony, of spores.   In all the
lower cryptogamia, the spores are such an important
factor, that all which concerns them collectively must
be of interest and importance, sufficient to condone
for a temporary digression.

FERNS, AND THEIR ALLIES.

FERNS are regarded as the highest in the scale of cryptogamic plants, and have always been great favourites. Perhaps none of the lower orders have been so much sought after, or so well investigated, so that their number may be estimated with something of certainty. In 1847 the total number of ferns, with their allies, was fixed by Lindley at 2264; but since that period the " Species Filicum," and other works, have augmented the number considerably, so that it cannot now fall much below 4000, and this, with 565 fern allies in 1887, would at least raise the total of ferns and fern allies to 4600 species. This estimate is based upon Hooker's and Baker's " Species Filicum," with the " Supplement," and Baker's " Fern Allies."

Although the ferns may have somewhat the appearance of flowering plants, it is by no means difficult for any ordinary person to distinguish them. In almost all cases it is sufficient to look at the under surface of the leaf or frond, when it will be observed that this under surface is not naked, as in an ordinary leaf, but furnished at regular distances with roundish or elongated brown elevated spots, or tufts, which a common pocket lens will demonstrate to be composed

of clusters of pear-shaped cases, attached by the thin
end, which split when fully ripe, and discharge a brown
dust consisting of the minute spores (Fig. 1).   These

are the representatives of the
flowers in flowering plants, and,
like them, are the organs of fructification.   The clusters of spore-cases
are termed *sori*, and are covered

Fig. 1.—Ferns, spore-case, at first, in most instances, with a
and spermatozoid.

thin membrane, but are occasion-
ally naked.   The form of the spore-cases, as well as
their position, varies in the different genera, but their
general character and function will be the same, that
of receptacles containing the spores, or representatives of seeds.   We need not say that these clusters
of spore-cases are very pretty and instructive objects
for the low powers of a microscope.

The above, then, will be sufficient for the determination of a fern, as distinguished from a flowering
plant, although there are other and less important
features which could be adduced in corroboration.
Suffice it to say that in the germination of the spores,
and the development of young plants, there are very
important differences from the process of reproduction
in flowering plants.   In common plants, when the
seeds germinate, a short stem is thrust out, bearing
one or two seed-leaves, which, although different in
form from the succeeding leaves, are true leaves.   At
the same time a small radicle, or rootlet, proceeds
downwards, and a young plant, resembling its parent,
has started into existence.   In ferns, on the contrary,
no such direct evolution takes place, but there is an

intermediary stage. When fern spores germinate,
a kind of filmy green membrane is produced, which
cannot be called a leaf; it lies flat upon the soil,
adhering by means of delicate rootlets proceeding
from its under surface, which, when mature, is lobed,
or almost heart-shaped. This delicate "false leaf"
is technically called a *prothallus*. Its structure and
its use is not that of an ordinary leaf, as it is only
a temporary expansion upon which the future young
fern is to be developed, and to call it a leaf would
simply be misleading. When this filmy "prothallus"
is matured, two kinds of organs are produced on its
under surface, and these represent the male and
female principle in the process of fertilization. One
kind of organ contains the germ of the future plant,
a kind of bud in its rudimentary state; this may be
termed the "germ-cell." The other organ represents
the fertilizing principle, and may roughly be designated
as a little bag or cell, containing active thread-like
bodies, which, when liberated from the cell, move
about as though endowed with animal life. When
the little bud or germ has reached a proper state of
maturity, it protrudes from its cell, and comes into
contact with some of these mysterious little active
thread-like bodies, and thus the bud is fertilized.
Gradually the lively bodies become still and disappear, the bud increases in size, the green membrane,
or prothallus, disappears also, and nothing is left but
the young bud, which begins to put on the appearance of a miniature fern plant, develops little fronds,
and grows up in the likeness of the parent plant
from whence the spores, or seeds, were first taken.

From the foregoing account it will be seen that the germination of a fern differs materially from that of ordinary plants, in that there is an intermediate generation. This alternation has been called an "alternation of generations," and something very like it occurs in the animal kingdom. The spore on germination produces a membranous prothallus. Secondly, the prothallus produces male and female organs, or fructification, by means of the germ-cells and the spermatozoids. From the union of these latter there proceeds a third generation, reverting to the original type, whilst the intermediate is different. Thus, then, we have a simple alternation of generations between the normal frondose type and the protothalloid, which latter is of but short duration. Something similar, but not so pronounced, takes place in the germination of mosses, when a confervoid protonema is developed, but of that anon.

The climatic relations under which ferns in general flourish, says Humboldt, "are manifested in the numerical laws of their quotients of distribution. In the plains within the tropical regions of large continents, this quotient is, according to Robert Brown, and other more recent investigations on the subject, $\frac{1}{20}$ of all phanerogamia, and in mountain districts of large continents $\frac{1}{6}$ to $\frac{1}{8}$. This ratio is quite different on the small islands scattered over the ocean; for here the proportion borne by the number of ferns to the sum total of all the phanerogamic plants increases so considerably, that in the South Sea Islands the quotient rises to $\frac{1}{4}$; while in the sporadic islands, St. Helena and Ascension, the number of ferns is almost equal to half of the whole phanerogamic vegetation. In receding from the tropics (where, on the large continents, D'Urville estimates the proportional number at $\frac{1}{10}$) the relative frequency of ferns decreases rapidly as we advance into the temperate zone. The quotients are for North America and the British Islands $\frac{1}{35}$, for France $\frac{1}{58}$, for Germany $\frac{1}{52}$, for the dry parts of Southern Italy $\frac{1}{74}$, for Greece $\frac{1}{84}$. The relative frequency again increases considerably towards the frigid north. Here the family of ferns decreases much more slowly in the number of its species than does that of phanerogamic plants. The luxuriantly aspiring character of the species, and the number of individuals contained in each, augment the deceptive impression of *absolute* frequency. According to Wahlenberg and Hornemann's catalogues, the relative numbers of the ferns are for Lapland $\frac{1}{25}$, for Iceland $\frac{1}{15}$, for Greenland $\frac{1}{12}$." [1]

Great favourites as are the ferns, especially for decorative purposes, they are, as a whole, singularly deficient in useful properties, only furnishing a kind of famine food, either to man or the lower animals, and not attractive for any special odour, or furnishing to the arts, medicine, or commerce, any product of importance.

TREE FERNS.

Ferns are theoretically plants without stems, with a terminal vegetation, the crown continually producing new fronds from its centre, whilst those at the circumference fade and decay. By this continual

[1] Humboldt, "Views of Nature," p. 340.

process, the adherent bases of the dead fronds gradually elevate the growing point, until a kind of spurious stem is formed, which in the "tree ferns" assumes the form and appearance of a trunk. Some of our common ferns, such as the "male fern," exhibit this process on a small scale, and from them an idea may be formed of the gradual growth of the pseudo-stem in tree ferns. When well grown, the magnificent tree ferns of the Southern Hemisphere vie with palms in elegance, and sometimes in size. New Zealand abounds in beautiful ferns. "Most conspicuous are the lofty and graceful arborescent ferns, of which there are several species. The 'Ponga' of the natives (*Cyathea dealbata*) is a noble tree; it grows abundantly on the declivities of the hills, under the shade afforded by the forests; it attains the height of fourteen or sixteen feet, crowned with its delicate fronds, which extend to a length of eight feet. Above, the fronds are of a fine dark green, but underneath of a beautiful silvery white colour. The circumference of the trunk is one foot and a half. Externally the trunk is composed of a black substance, hard as ebony, which is continued into the interior, intersecting the white medullary part. When the tree is cut down, an adhesive juice exudes from it. The natives use the trunk of this fern as posts, in the erection of their dwellings, and they are very durable—the medullary portion soon decaying, but the exterior lasting for several years. There are two other species, surpassing in magnificence of growth that just mentioned. I accompanied a native to a place where I could observe them

FIG. 5.—Tree Ferns. From Gosse's "Romance of Natural History."

growing. After passing through a dense forest, we descended a hill covered with exuberant vegetation, and shaded by enormous trees ; we then came upon a marshy spot, luxuriant in vegetation, where the magnificent tree ferns (*Dicksonia*) rose in clumps before us. Solitude reigned, only disturbed by the low murmuring of the silver rivulets, as they meandered through the richly verdant banks. The largest of these magnificent ferns is about twenty feet high, and the trunk two feet in circumference. It is remarkable from the large size of the spiral stipes, and the enormous extent of its fronds ; the trunk, stipes, and central stalks of the fronds are of a beautiful shining black colour ; the length of the fronds is from sixteen to eighteen feet, and the leaflets two to three feet. Not far distant grew the other species, which attains the size of the Ponga both in trunk and extent of fronds, but the leaflets are smaller, and the stalk and under surface of the fronds are yellow. These two species thrive in marshy ground, and in dense, shady localities." [1]

From another source we learn that the "silver fern," or *Ponga*, reaches far greater dimensions than those stated above, since the trunk will acquire a height of forty feet, and it is peculiar to New Zealand, where another indigenous species reaches a height of twenty-four feet. In Tasmania the beautiful *Dicksonia* grows to from thirty to fifty feet in height, with sometimes a diameter of not less than four feet, and fronds twelve feet in length. Another

---

[1] George Bennett, M.D., " Gatherings of a Naturalist in Australasia," p. 417. (1860.)

species, *Alsophila*, attains a height of thirty feet, with a rough black trunk, and fronds of from twelve to fourteen feet in length.   The giant of tree ferns, however, is found in Norfolk Island, where *Alsophila excelsa* will grow to the height of eighty feet.

An experienced traveller, visiting the Aru Islands, declares that "the greatest novelty and most striking feature to his eyes were the tree ferns, which, after seven years spent in the tropics, he then saw in perfection for the first time.   All he had hitherto met with were slender species, not more than twelve feet high, and they gave not the slightest idea of the supreme beauty of trees, bearing their elegant heads of fronds, more than thirty feet in the air, like those which were plentifully scattered about this forest. There is nothing in tropical vegetation so perfectly beautiful." [1]

Dr. J. D. Hooker says, in his "Himalayan Journals," that "the most interesting botanical ramble about Silhet is to the tree-fern groves on the path to Jyntespore, following the bottoms of the shallow valleys, and along clear streams, up whose beds we waded for some miles, under an arching canopy of tropical shrubs, trees, and climbers.   In the narrower parts of the valleys tree ferns are numerous on the slopes, rearing their slender brown trunks forty feet high, with feathery crowns of foliage, through which the sunbeams trembled on the broad shining foliage of the tropical herbage below." [2]

Humboldt has remarked it as singular that no

---

[1] A. R. Wallace, " The Malay Archipelago," vol. ii. p. 209.   (1869.)
[2] Hooker, " Himalayan Journals," vol. i. p. 325.

C

mention of the beautiful tree ferns is to be found in
the classic authors of antiquity, Theophrastus, Dios-
corides, and Pliny ; while reference is made to the
Indian trees, to the fig-tree which takes root from its
branches, and to palms.   The first mention of tree
ferns is by Oviedo (1535).   "Among ferns," says this
experienced traveller, " there are some which I class
with trees, because they are as thick and high as
pine-trees.   They mostly grow among the mountains,
and where there is much water."   After complaining
that the height is exaggerated, Humboldt goes on to
say that " in the *Cyathea speciosa* and the Meniscium
of the Chaymas missions he observed in the midst of
the most shady part of the primeval forest that the
scaly stems of some of the most luxuriantly developed
of these trees were covered with a shining carbo-
naceous powder, which appeared to be owing to a
singular decomposition of the fibrous part of the old
leaf stalks." [1]

Between the tropics, where, on the declivities of the
Cordilleras, climates are superimposed in strata, the
true region of arborescent ferns lies between about
3200 and 5350 feet above the level of the sea.   In
South America and in the Mexican highlands they
seldom descend lower to the plains than 1280 feet.
The mean temperature of this happy region is between
64° and 70° Fahrenheit.   It reaches the lowest stratum
of clouds, which floats the nearest to the surface of
the sea and the plain, and it therefore enjoys unin-
terruptedly a high degree of humidity, together with
a great equality in its thermal relations.

[1] Humboldt, " Views of Nature," p. 338.

The conditions of genial mildness in an atmosphere charged with aqueous vapour, and of great uniformity in respect to moisture and warmth, are fulfilled on the declivities of the mountains in the valleys of the Andes, and more especially in the southern milder and more humid atmosphere, where arborescent ferns advance not only to New Zealand and Tasmania, but even as far as the Straits of Magellan and Campbell Island, and therefore to a southern latitude almost identical in degrees with the parallel in which Berlin is situated north of the equator.

All travellers who have visited the "fern gullies" in Australasia, speak rapturously of the impression made upon them by a first sight of a grove of these "kings and princes" of cryptogamic vegetation. The trunks may not be so lofty as those of the palms, but are more picturesque from their rugged surface. The fronds are of a brighter and more vivid green, and far more elegant in their tracery, and the delicate sub-divisions of their compound pinnules. What they may lose in majesty they gain in beauty, and, not being so far removed from the eye, all the advantages are increased in effect.

SCYTHIAN LAMB.

The old story of the "Scythian" or "Tartarian Lamb" has been referred to a species of tree fern, or *Cibotium*. Darwin, in his "Loves of the Plants," describes the mystic object as generally represented in folk-lore—

> " Cradled in snow, and fanned by Arctic air,
> Shines, gentle Barometz ! thy golden hair
> Rooted in earth, each cloven hoof descends,
> And round and round her flexile neck she bends ;
> Crops the gray coral-moss and hoary thyme,
> Or laps with rosy tongue the melting rime,
> Eyes with mute tenderness her distant dam,
> Or seems to bleat, a vegetable lamb."

In the frontispiece to Parkinson's " Paradisus " the Barometz, or vegetable lamb, is represented as one of the plants growing in Eden. The woolly body of the lamb, with its golden hair, is supposed to be the hairy rhizome, or bases of the stipes of *Cibotium Barometz*, the legs being the stipes of four fronds. The figure given by Zahn (1696) is by no means like one we have met with in some old work, which has been introduced, reduced in size, into the " Treasury of Botany." The translation of Zahn's remarks runs as follows : " Very wonderful is the Tartarian shrub or plant, which the natives calls *Baromez*, i.e. Lamb. It grows like a lamb, to about the height of three feet. It resembles a lamb in feet, in hoofs, in ears, and in the whole head, save the horns. For horns, it possesses tufts of hair, resembling a horn in appearance. It is covered with the thinnest bark, which is taken off and used by the inhabitants for the protection of their heads. They say that the inner pulp resembles lobster flesh, and that blood flows from it when it is wounded. Its root projects and rises to the umbilicus. What renders the wonder more remarkable, is the fact that, when the Baromez is surrounded by abundant herbage, it lives as long as a lamb in pleasant pastures ; but when they become exhausted, it wastes away and perishes. It is said that wolves have a liking for it, while other carnivorous animals have not." [1]

Scaliger gives a similar description, adding that it is not the fruit, the melon, but the whole plant that resembles a lamb. This does not tally with the account given by Odorico da Pordenone, an Indian traveller, who, before the Barometz had been heard of in Europe, was informed that a plant grew on some island in the Caspian Sea which bore melon-like fruit, resembling a lamb ; and this tree is figured and described by Sir John Maunderville, with the young lambs projecting from the fruits.

The Barometz, or Scythian Lamb (*Cibotium Barometz*), is the name given to a fern growing in Tartary, the root of which, says Professor Martyn, from the variety of its form, is easily made by art to take the form of a lamb (called by the Tartars *Barometz*), " or rather that of a rufous dog, which the common names in China and Cochin China imply, namely *Cau-tich*, and *Kew-tsie*." The description given of this fern represents the root as rising above the ground in an oblong form, covered all over with hairs ; towards one end it frequently becomes narrower, and then thicker, so as to give somewhat of the shape of a head and neck, and it has sometimes two pendulous hairy excrescences resembling ears ; at the other end a short shoot extends out into a tail. Four fronds are chosen in a suitable position, and are cut off to a proper length to represent the legs ; and thus a vegetable lamb is produced.[2]

[1] Richard Folkard, Jun., "Plant-lore, Legends, and Lyrics," p. 121. (1884.)   [2] Ibid , p. 243.

Several of the larger ferns have the stipes clad with shining brown hairs, and this substance, stripped from the stipes, has entered into commerce under the name of *Penawar jambie*, being used, not only for stuffing cushions, but also as a styptic. The brief description of this substance is to the effect that " it is the shining moniliform silky hairs of the rhizome

Fig. 3.—Hairs of the rhizome of *Cibotium Barometz* (J. Smith), from a specimen cultivated in the Royal Gardens, Kew (magnified about 30 diameters). (From *Pharmaceutical Journal.*)

and stipes of *Cibotium Barometz*, and is produced in Sumatra. It is the ' Scythian Lamb ' of the old writers." [1] The fern makes a little stem of about one foot in height, and fronds of five feet in length. The down appears on the young undeveloped leaves, and

[1] *Pharmaceutical Journal*, April, 1860.

as these grow it remains upon the stem, between the
leaf stalks (Fig. 3). Old stems bear a great quantity
of this down. The native female physicians pick for
years at such a stem without exhausting the stock.

The produce of another fern is called *Pakoe kidang*,
of which a plantation exists on the sides of the
Goenong Gedeh, a volcano in the interior of Java,

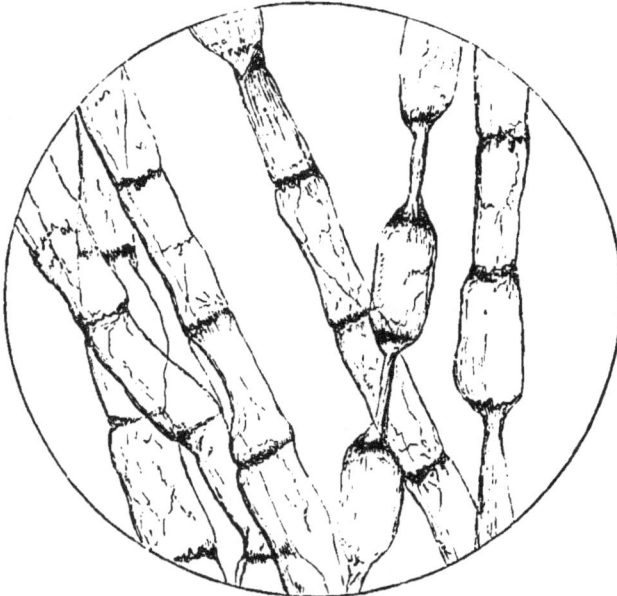

FIG. 4.—Larger hairs of *Pakoe Kidang* (magnified about 30 diameters). (From
Pharmaceutical Journal.)

between sixty and seventy miles from Batavia. The
brown hairy down of the stipes in this species (*Balantium chrysotrichum*) is coarser than in the foregoing,
and less silky (Fig. 4). To these must be added the
substance called *pulu*, which is solely the produce of the
Sandwich Islands, derived from one or more species
of *Cibotium*. It was introduced into Liverpool fifty

years ago, as a commercial article, for stuffing pillows, etc.; and in 1858 no less than some 313,000 lbs. were exported from the Sandwich Islands, and the trade was gradually increasing. Owing to the large quantities collected, it was then becoming more rare, and the price advanced yearly. " The number of persons engaged in gathering *pulu* varies. Including men, women, and children, probably from two to three thousand are now dependent on it for a livelihood, receiving generally from five to six cents per pound on delivery. The labour of gathering pulu is very tedious and slow. When picked, it is wet, and has to be laid out to dry on the rocks or on mats. In favourable weather it will dry in a day or two, but generally in the pulu region wet and rainy days prevail, so that frequently the natives do not get their pulu dry after several weeks, often taking it to market in too wet a state. The dealers have constantly to contend with this inclination of the natives to sell wet pulu, as it makes considerable difference in the weight when dry. The facilities for drying, packing, and shipping are improving every year, and the article now shipped is generally dry and in good order, closely packed in wool bales. The trade is reduced to a system, and though there is no probability of any great increase, it will doubtless continue a staple export."[1]

A similar substance, which is the produce of *Balantium culcita*, in Madeira and the Azores, is used for stuffing cushions, mattresses, etc. Indeed, this list might be considerably extended, but we are princi-

---

[1] M. C. Cooke, "On Pulu," in *Pharmaceutical Journal*, April, 1860.

pally interested in the hairs of the Scythian Lamb, and to the others we have alluded incidentally. The romance of one age becomes the servant of the next.

TARA FERN.

It is only in rare cases, and those almost of necessity, that the natives in various countries resort to ferns for supplying an article of food. An exceptional instance is reported by Mr. James Backhouse, of the use of the Tara Fern in Tasmania, and he states that "the most extensively diffused eatable roots of Van Dieman's Land are those of the Tara Fern," which greatly resembles the common Bracken, covering great extents of light and rich land. The botanical name is *Pteris esculenta.* "It varies in height from a few inches to several feet, according to the richness of the soil, and in some parts of the colony is so tall as to conceal a man on horseback. The root is not bulbous, but creeps horizontally at a few inches below the surface of the earth, and when luxuriant attains the thickness of a man's thumb. Pigs feed upon this root where it has been turned up by the plough, and in sandy soils they will themselves grub up the earth in search of it. The aborigines roast it in the ashes, peeling off its black skin with their teeth, and eating it as sauce to their roasted kangaroo, in the same manner as Europeans use bread. This root possesses much nutritive matter, yet it is observed that persons who have been reduced to the use of it, in long excursions through the bush, have become very weak, though it has supported life. Whether this arose from an insufficient supply, in consequence of the parties being too much exhausted to dig it up before they resorted to it, or from eating it raw, or some other cause, I am not able to determine. It is quite certain that when this substance is grated, or reduced to a pulp by beating and mixing with cold water, a large quantity of arrowroot is precipitated, which adheres to the bottom of the vessel, and which may easily be prepared for use by pouring off the water and floating matter, adding fresh water, and stirring up the white powder, and again allowing it to settle. It may then be cooked by boiling, or the powder may be spread on cloths and dried in the sun, or hung up in linen bags where there is a free circulation of the air." [1]

It is interesting to note that the common Bracken has sometimes been used in Northern Europe and Siberia to make a coarse kind of bread. The rhizome, or underground stem, is the part employed for that purpose, as in the case of the Tara Fern. Having a knowledge of this application of fern root, the Rev. M. J. Berkeley, in 1856, made some experiments in order to ascertain what kind of food it would furnish. " I accordingly roasted some of the rhizomata (underground stem, popularly called the root) and found them eatable, but extremely disagreeable from their slimy consistence and peculiar flavour, in both of which respects they precisely resemble ill-ripened Brinjals. It struck me, however, that they might afford a better food if the slimy matter could be removed. I accordingly scraped some of the rhizomata,

[1] On the esculent plants of Van Dieman's Land, in " Companion to Botanical Magazine," vol. ii. p. 39. (1836.)

which had first been washed and peeled, avoiding, however, the two columns of hard coloured tissue with which they are threaded, and then placed the pulp thus obtained in water. After four and twenty hours, the water had become extremely slimy and of a yellow brown. This was carefully decanted, and the pulp washed again with water, which was now quite colourless. This also was decanted, and the pulp, when sufficiently dry, was kneaded into a cake and baked upon the hearth. The result was a coarse but palatable food, perfectly free from any disagreeable flavour—much better, indeed, to my taste, and probably not less nutritious than Cassava bread."[1]

After all, we may rest fairly well assured that ferns and lichens only furnish a kind of famine food to be employed in case of dire necessity, otherwise their consumption would not be so limited. When it is taken into account how much labour must be expended in collecting and preparing fern root for food, which, at its best, is only to be described as palatable, we may rest content that it is not worth the trouble to make the experiment, except to gratify curiosity. Supposing the edible portion to consist of starch or lichenin, or some equivalent, the quantity must be very small in proportion to the original bulk, and that of an inferior quality.

Another fern is also utilized by the native blacks of Van Diemen's Land, who split open about a foot and a half of the top of the trunk of the common Tree Fern, *Cibotium Billardieri*, and, taking out the heart,

---

[1] Rev. M. J. Berkeley, in *Journal of Linnæan Society*, 1857, vol. i. p. 156.

which resembles a Swedish turnip in substance and is as thick as a man's arm, they roast it in the ashes and eat it like bread ; but it is too bitter and astringent to suit an English palate. It is said also that the aborigines eat, and prefer, the heart of another tree fern, *Alsophila australis*, which is found on the western side. We have tasted of a kind of bread which the aborigines concoct from the pollen of the Reed Mace (*Typha*), and, if that may be accepted as a type of what the aborigines consider an enjoyable food, we are content with the seeds of the cereals, even those of the poorest kind, rather than adopt the manners and customs of the aborigines. The romantic stories are sufficient to satisfy our curiosity without resorting to practical demonstration. An old mycophagist may be excused a little prejudice, but, in his opinion, there are really no known cryptogams which furnish a desirable article of food, except fungi, and of them no true epicure can ever tire.

FERN SEED.

One little romance associated with ferns is that the minute spores, sometimes called seeds, if gathered under peculiar circumstances, would render the possessor invisible. Old Gerard says that " fern is one of those plants which have their seed on the back of the leaf, so small as to escape the sight. Those who perceived that fern was propagated by semination, and yet could never see the seed, were much at a loss for a solution of the difficulty ; and, as wonder always endeavours to augment itself, they

ascribed to fern seed many strange properties, some
of which the rustick virgins have not yet forgotten, or
exploded." Ben Jonson alludes to the invisibility
of those who carry fern seed, thus :—

> " I had
> No medicine, sir, to go invisible,
> No fern seed in my pocket."

Other properties are also ascribed to fern seed,
which were taken into account in the operations of
Midsummer Eve. "Fern seed," says Grose, "is
looked on as having great magical powers, and must
be gathered on Midsummer Eve. A person who
went to gather it reported that the spirits whisked
by his ears, and sometimes struck his hat, and other
parts of his body ; and, at length, when he thought
he had got a good quantity of it, and secured it in
papers and a box, when he came home he found
both empty." Torreblanca, in his " Demonology,"
suspects those persons of witchcraft who gather fern
seed on this night. Brand says, " A respectable
countryman at Heston, in Middlesex, informed me,
in June, 1793, that, when he was a young man, he
was often present at the ceremony of catching the
fern seed at midnight, on the eve of St. John
Baptist. The attempt, he said, was often unsuccessful,
for the seed was to fall into the plate of its own
accord, and that, too, without shaking the plant."

In an old translation of Pliny it is written, "of
ferne be two kinds, and they beare neither floure nor
seed." The ancients, who often paid more attention
to received opinions than to the evidence of their
senses, believed that fern bore no seed. Our ancestors

imagined that this plant produced seed which was invisible. Hence, from an extraordinary mode of reasoning, founded on the fantastic doctrine of signatures, they concluded that they who possessed the secret of wearing this seed about them would become invisible.[1]

This belief is alluded to by Beaumont and Fletcher, in *Fair Maid of the Inn.*

> " Why, did you think that you had Gyges' ring,
> Or the herb that gives invisibility ? "

And Shakespeare, in *Henry IV.*, part i., act ii., scene i.—

> " *Gadshill.* We have the receipt of fern seed, we walk invisible.
> " *Chamberlain.* Nay, by my faith ; I think rather you are more beholding to the night than to fern seed, for your walking invisible."

To catch the wonder-working seed, twelve pewter plates must be taken to the spot where the fern grows ; the seed, it is affirmed, will pass through eleven of the plates, and rest upon the twelfth. This is one account ; another says that midsummer night is the most propitious time to procure the mystic fern seed, but that the seeker must go barefooted, and in his shirt, and be in a religious state of mind.

De Gubernatis gives more explicit instructions as obtained from a Russian peasant. " On midsummer night, before twelve o'clock, with a white napkin, a cross, a Testament, a glass of water, and a watch, one seeks in the forest the spot where the fern grows ; one traces with the cross a large circle ; one spreads the napkin, placing the cross on the Testament, and

[1] Brand's " Popular Antiquities," vol. i. p. 315. (1849.)

the glass of water. Then one attentively looks at one's watch; at the precise midnight hour the fern will bloom; one watches attentively, for he who shall see the fern seed drop shall at the same time see many other marvels; for example, three suns and a full moon, which reveals every object, even the most hidden. One hears laughter, one is conscious of being called; if one remains quiet, one will hear all that is happening in the world, and all that is going to happen."[1]

Another writer, Markevic, also says,[2] "The fern flowers on midsummer night at twelve o'clock, and drives away all unclean spirits. First of all it puts forth buds, which afterwards expand, then open, and finally change into flowers of a dark red hue. At midnight the flower opens to its fullest extent, and illuminates everything around. But at that precise moment, a demon plucks it from its stalk. Whoever wishes to procure this flower must be in the forest before midnight, locate himself near the fern, and trace a circle around it. When the evil spirit approaches, and calls, feigning the voice of a parent, sweetheart, etc., no attention must be paid, nor must the head be turned, for if it is, it will remain so. Whoever becomes the happy possessor of the flower has nothing to fear; by its means he can recover lost treasure, become invisible, rule on earth, and under water, and defy evil spirits. To discover hidden

[1] De Gubernatis, "La Mythologie des Plantes; ou les Legends du Regne Végétale." (Paris.)

[2] Richard Folkard, Jun., "Plant Lore, Legends, and Lyrics," p. 332. (London, 1884.)

treasure it is only necessary to throw the flower in the air ; if it turns like a star above the sun, so that it falls perpendicularly in the same spot, it is a sure indication that treasure is concealed there." Certainly it is needless to add that, as ferns are flowerless plants, the flower of the fern is in itself a myth.

Folkard states that "the people of Westphalia are wont to relate how one of their countrymen chanced one midsummer night to be looking for a foal he had lost, and passing through a meadow, just as the fern seed was ripening, some of it fell in his shoes. In the morning, he went home, walked into the sitting-room, and sat down, but thought it strange that neither his wife, nor indeed any of the family, took the slightest notice of him. ' I have not found the foal,' said he. Everybody in the room started, and gazed around, with scared looks, for they had heard a man's voice but saw no one. Thinking that he was joking, and had hid himself, his wife called him by his name. Thereupon he stood up, planted himself in the middle of the floor, and said, ' Why do you call me ? Here I am, right before you.' Then they were more frightened than ever, for they had heard him stand up and walk, and still they could not see him. The man now became aware that he was invisible, and a thought struck him, that possibly he might have got fern seed in his shoes, for he felt as if there was sand in them. So he took them off, and shook out the fern seed, and as he did so he became visible to everybody."

The English tradition is that the fern blooms and seeds only at twelve o'clock, on St. John's Eve, just at the precise moment at which John the Baptist was born. In Dr. Jackson's works (1673) we read that he once questioned one of his parishioners as to what he saw or heard when he watched the falling of the fern seed, whereupon the man informed him that this good seed is in the keeping of Oberon, King of the Fairies, who would never harm any one watching it. He then said to the worthy doctor, "Sir, you are a scholar, and I am none. Tell me, what said the angel to our Lady; or what conference had our Lady with her cousin Elisabeth, concerning the birth of St. John the Baptist?" Finding Dr. Jackson unable to answer him, he told him that the angel did foretell John Baptist should be born at that very instant in which the fern seed—at other times invisible —did fall; intimating further that this saint of God had some extraordinary virtue from the time or circumstance of his birth.

Enough has been said to show the kind of romance which was current amongst illiterate and superstitious people with regard to ferns, and especially as to the dispersion of the spores, or seed, some two hundred years ago, the relics of which superstition even extended to much later times. The freedom with which ferns are planted in the gardens, and cherished in the domiciles, of "all sorts and conditions of men," lead to the conclusion that no suspicions lurk in the minds of the most uneducated that they are unlucky, albeit it is stated that in olden times, when witches mounted the clouds and rode the winds on broomsticks, it was the Moon-fern which made the saddle of their fleet steeds.

D

NARDOO.

There is a melancholy interest connected with the Nardoo plant of Australia, on account of its association with the unfortunate explorers who were starved on the homeward journey, after traversing the continent of Australia. The passage occurs in Wills's "Diary," where reference is made to the party falling in with some blacks who were fishing. "They gave us some half a dozen fish each for luncheon, and intimated that if we would go to their camp we should have some more, and some bread. On our arrival at the camp, they led us to a spot to camp on, and soon afterwards brought a lot of fish and bread, which they called *nardoo*. In the evening various members of the tribe came down with lumps of nardoo, and handfuls of fish, until we were positively unable to eat any more."[1] The plant from which this substance was procured has been determined as a species of *Marsilea*, a genus allied to the ferns, the sporocarps, or fruits, of which are pounded into a kind of flour. Some of the fruits were brought away by Mr. King, the survivor of the party.

About forty species of *Marsilea* are known, and this one has had several names applied to it, having settled down, at length, to *Marsilea Drummondi*. They hardly resemble ferns in appearance, and the present is almost aquatic. The petiole of the fronds, or leaves, is about six inches long, and slender, crowned with a four-leaved frond, like a trefoil with

---

[1] Andrew Jackson, "Account of the Expedition" (Smith and Elder, 1862); "On the Nardoo Plant," F. Currey, *Seem. Journ. Bot.* (1863), i. 161.

four leaflets, or the traditional "four-leaved shamrock."
The conceptacles are somewhat of the character of
those in ferns, and contain sporangia of two kinds,
that is, micro-sporangia containing numerous micro-
spores, and macro-sporangia with solitary macrospores.
The latter are almost half-moon shaped, obtuse,
entirely bald, furnished with many manifest ribs, with
two short teeth at the suture.

The sporocarps of the "Nardoo" obtained by Mr.
King were sent to Dr. Moore of Glasnevin for culture,
and the results of the examination and experiments
were afterwards published. He says that "the fact
of the fruit of any cryptogamous plant containing a
sufficient quantity of nutritive substance to support
human life during a lengthened period, at once struck
me as being a very remarkable circumstance. It has
been long known that the thallus of some, and the
rhizomes of others, contain nutritive matter, which
leads to their being occasionally used as food by the
natives of various parts of the world, but this I believe
to be the only instance on record of the fruit of any
of them being employed for that purpose. Poor
Burke and his companions were able to subsist on
this during a considerable period, but they also died
on it, with the exception of King, who was reduced
to a mere skeleton when found by the relief party.
I believe that I am pretty safe in assuming that the
nutritive properties contained in the thallus, or
rhizomes, of cryptogamic plants depend chiefly on the
presence of an amylaceous substance, analogous to
gelatine, which occurs in the form of pure starch, or
amylaceous fibre, which is also the case in the fruit

of the 'Nardoo.' Afterwards," he adds, "the body which germinates, and produces the future plant, is filled with well-defined and very large starch granules. I applied the test of iodine to them, which speedily turned them a violet-blue colour, thus revealing their true nature, and at the same time affording evidence of the principal source of nutrition in the 'Nardoo.'"[1] It appears from King's narrative, that the preparation consists of pounding the fruits between stones, and baking into cakes, as we use flour, or simply boiling.

The "Nardoo" must be classed with the "Tripe de Roche" as famine plants, which have been resorted to in dire extremity, and which for a time have sustained human life. As such they are to be remembered, although no one would think of resorting to them, as articles of food, unless impelled thereto by necessity.

Club Moss.

The club mosses are a small genus of plants, of which we possess some six British species, having much the habit of large mosses, but with a reproduction more closely allied to the horsetails and ferns. In some places they are called Stag's-horn or Fox's-tail, and are known, when the fruit is ripe, to contain a quantity of fine dusty powder, which shakes out easily, and was, in former times, more in use than at present. This powder is highly inflammable, and is the "lightning meal" of the Germans, because employed in the manufacture of fireworks, and at one

---

[1] "On the Nardoo Plant," by David Moore, in *Gardener's Chronicle*, Aug. 30, 1862, p. 812.

time for the production of stage lightning. Under the name of *lycopodium* it was employed also for the coating of pills, but is now almost excluded from Materia Medica.

One of the species is *Lycopodium selago*, which is believed to be the plant held in so great a repute by the Druids. It was called Golden Herb, and was reputed to confer the power of understanding the language of birds and beasts. Pliny says that the Druidic Selago resembled Savin, and that it was gathered, as it by stealth, without the use of iron. "The person who gathers it must go barefoot, with feet washed, clad in

FIG. 5.—*Lycopodium*, or Club Moss.

white, having previously offered a sacrifice of bread and wine, and must pluck the plant with his right hand, through the left sleeve of his tunic. It is carried in a new cloth."[1] Old Gerard said that the club moss, or Heath Cypress, was thought to be the *Selago* mentioned by Pliny. "The catkins of this plant are described as being of a yellowish colour, and it is stated to be found growing in divers woody

[1] Pliny, "Natural History," xxiv. 62.

mountainous places of Germany, 'where they call it
Wilde Savine." In Reade's "Veil of Isis" is a similar
account of the gathering of the Selago, excepting
that it was cut with a brazen hook.

Sir W. J. Hooker has given a summary of the
properties and uses of the club moss, and its products,
which leaves nothing to be desired. "The whole
plant," he says, alluding to *Lycopodium clavatum*,
"possesses peculiar qualities," but is most celebrated
for the yellowish inflammable and detonating dust,
which even resembles gunpowder in the two latter
respects, and is afforded by its capsules in an immense
quantity. This substance is largely collected, and
applied to different purposes, being known by the
vulgar name of Vegetable brimstone, or Lycopode.
A pinch of it, when cast upon any burning matter,
takes flame instantly, darting forth a blaze which
almost immediately disappears, and without leaving
any perceptible odour. It is this singular property
which has caused the lycopode to be employed on
the stage to represent lightning, infernal flames, etc.,
as well as in the preparation of fireworks. Its
consumption is so great as to render it a rather
lucrative object of commerce in Switzerland and
Germany, where this vegetable powder is principally
collected, and where it is often adulterated with the
staminal dust of the fir-cones, which, however, possesses
none of its qualities. Towards the close of summer,
during autumn, and the commencement of winter,
the spikes of this *Lycopodium* appear, and diffuse the
lycopode contained in their capsules (Fig. 5). They
are cut off and carried home to be dried in boxes or

sieves prepared for the purpose; and being shaken from time to time, the powder drops out, when it is collected and, after being dried anew, is fit for sale. In pharmacy this dust is used to roll up boluses and pills, the result being to cover them with a foreign substance which preserves them unaltered. In fact, so completely does the lycopode coat the surface of the pills, that they may be put into water, and taken out again, without being moistened, an experiment which may be still more satisfactorily made by putting one's hand into water into which lycopode has been thrown, when the hand will come out dry. The adherence of these minute seeds to one another, is doubtless the cause of this phenomenon." [1]

[1] W. J. Hooker, "Of Lycopodium," in *Annals of Natural History*, Aug., 1838, p. 428.

MOSSES.

THESE pretty little flowerless plants are popularly great favourites, and deservedly so, but
deficient in very romantic associations, or remarkable
life-history.   These are so well known in general
appearance that, fortunately, general description is
unnecessary.   The bryologist—that is, the person
who studies mosses—recognizes, in addition to the
true mosses, two other small groups, viz. bog or peat
mosses, and split mosses, or *Andreaceæ*, which are
included, for our present purpose, in a broad sense,
as mosses.

Going back to Lindley's "Vegetable Kingdom" of
1847, we find the estimate for all known species at
that time to have been about 1120; whereas British
species alone are now more than half that number.
A friend, who is addicted to the study of mosses,
informs us that he considers eight thousand species
to be below the total up to date, and that possibly
they might reach to nearly ten thousand for the
entire world, as at present known.

It may be premised that mosses are cellular plants
having a distinct stem, furnished with leaves, and
prolonged downwards into a root, with the exception

of the peat mosses. Their reproduction is accomplished either by spores or by budding; as a few
species never produce fruit in this country, their
extension and preservation is accomplished by the
process of budding, or gemmation. Vegetation proceeds in the same manner as in flowering plants, in
so far as it is terminal, progressing by gradual elongation of the apex, or grow-
ing point. The stem is
therefore very variable in
length; in some species
being invariably short, and
almost imperceptible, but
in others considerably extended, even to a length of
several inches. One of the
most interesting features
in the development of
mosses is the fructification,
for although called " flowerless plants," they really
possess the analogues of
male and female flowers.
The male flowers are usually
termed Antheridia, and the
female Archegonia, the
former containing the male
element for the purpose of
fertilization, and the latter
the ovary, enclosing the
ovules, to be matured into spores. For the minute
details of these structures, we must refer to some

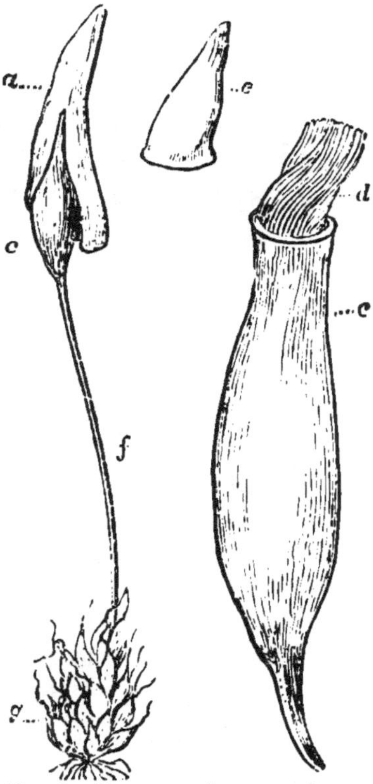

Fig. 6.—Moss capsule. *a*, calyptra;
c, theca, or capsule; *f*, fruit stalk;
g, leaves; *d*, peristome; *e*, conical
lid.

manual,[1] as their description here would extend to
too great a length. The mature fruit consists of an
elegant little capsule (Fig. 6), which opens at the
apex by an operculum, or lid (*e*). Between the edge
of the capsule and the lid there is usually a ring of
minute cells, which assist in casting off the lid when
the spores are mature. In most of the genera the
mouth is surrounded by a fringe of delicate teeth,
always some multiple of four, which is a most enter-
taining object under the microscope (*d*). In the
Andreaceæ the capsule does not open with a lid, but
splits into four valves.

The spores are minute rounded bodies, produced

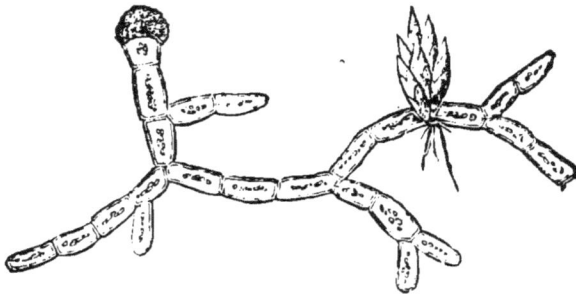

FIG. 7.—Protonema of a moss.

in considerable number within the capsules, readily
dispersed, when mature, by the falling away of the
lid, or operculum. When the spores germinate, which
they do from any portion of their surface, they produce
a green filamentous body called a *protonema* (Fig. 7),
which is a kind of prothallus, often branched and
much elongated. In this condition it has a close

---

[1] Consult especially a "Handbook of British Mosses," by J. E.
Bagnall, A.L.S.

resemblance to some forms of the fresh-water Algæ. Subsequently little buds arise on various parts of the protonema, which send down rootlets into the matrix, and become rudimentary moss plants. Thereafter progress is normal, by the production of stem and leaves, and they partake of a perfect resemblance to the original moss from which the spore was derived.

This intercalation of a more or less thalloid intermediary between the spore and young plant has somewhat of an analogue throughout most of the cryptogamia. Its greatest perfection is in the ferns, where the prothallus develops sexual organs, as a true alternation of generations ; sinking in the mosses to a protonema, on which young plants arise by budding. In the lichens these are represented by the thallus, which bears the apothecia, and sometimes remains long as a barren thallus, including the whole vegetative system of the plant. But in fungi it becomes degraded into mere threads of mycelium, or consolidated into hybernating sclerotia. Lastly, in the Algæ it may be normally obsolete, or, represented by active zoospores, eventuate in a combination of vegetative with reproductive systems. Thus, then, in all we may recognize a middle state between the mature ovum and the young plant, which may be either a prothallus, a protonema, a thallus, a mycelium, filamentous or compact, or a zoospore. Some may regard this as a fanciful analogy, but, if so, it will not jeopardize the facts, which will remain stable amid the ruins of the hypothesis.

PEAT MOSS.

The peat mosses, or bog mosses, have long been recognized as possessing features which entitle them to consideration apart from the frondose mosses, with which they have sometimes been associated. The peat mosses, known also as *Sphagnaceæ*, by their structure and habits constitute a natural and compact group, which may well be treated by itself, and is of sufficient interest to receive a brief reference in a work of the present character. Dr. R. Braithwaite has rightly observed that " few persons can have traversed our moorlands without having had their attention attracted to the great masses of *Sphagnum* (Fig. 8) which adorn their surface—now in dense cushions of lively red, now covering some shallow pool with a vast sheet of light green, inviting, it may be, by its bright colour ; but woe betide the inexperienced collector who sets foot thereon, for the spongy mass may be many feet in depth, and he may run the chance of never reaching *terra firma* again." [1]   We need not stay to inquire into the minute structure and characteristics of this little group.   Nor is it

Fig. 8.—Squarrose Bog Moss.

[1] Braithwaite, "The Sphagnaceæ, or Peat Mosses of Europe and North America," p. 6.   (1880.)

incumbent in us to show wherein they differ from mosses generally. It is sufficient to say that the only apology for special allusion to them is on account of their share in the composition of those peat beds from which their name is derived. Professor Schimper has expressed all this in a few lines. He says, " Unless there were peat mosses, many a bare mountain ridge, many a high valley of the temperate zone, and large tracts of the northern plains, would present a uniform watery flat, instead of a covering of flowering plants or shady woods. For just as the *Sphagna* suck up the atmospheric moisture, and convey it to the earth, do they also contribute to it by pumping up to the surface of the tufts formed by them the standing water, which was their cradle, diminish it by promoting evaporation, and finally also, by their own detritus, and by that of the numerous other bog-plants to which they serve as a support, remove it entirely, and thus bring about their own destruction. Then, as soon as the plant detritus formed in this manner has elevated itself above the surface water, it is familiar to us by the name of *peat*, becomes material for fuel, and all *Sphagnum* vegetation ceases."

MUNGO PARK'S MOSS.

Mosses have not, as a rule, played any considerable part in the romance of the lower forms of vegetable life, but the incident, so often quoted, in connection with the traveller, Mungo Park, cannot be omitted. This enterprising traveller, during one of his journeys into the interior of Africa, was cruelly stripped and

robbed of all he possessed by banditti. "In this for-
lorn and almost helpless condition," he says, "when
the robbers had left me, I sat for some time, looking
around me with amazement and terror. Whichever
way I turned nothing appeared but danger and difficulty. I found myself in the midst of a vast wilder-
ness, in the depth of the rainy season—naked and
alone—surrounded by savage animals, and by men
still more savage. I was five hundred miles from
any European settlement. All these circumstances
crowded at once upon my recollection, and I confess
that my spirits began to fail me. I considered my
fate as certain, and that I had no alternative but to
lie down and perish. The influence of religion, however, aided and supported me. I reflected that no
human prudence or foresight could possibly have
averted my present sufferings. I was indeed a
stranger in a strange land, yet I was still under the
protecting eye of that Providence Who has con-
descended to call Himself the stranger's Friend. At
this moment, painful as my reflections were, the
extraordinary beauty of a small moss irresistibly
caught my eye; and though the whole plant was
not larger than the top of one of my fingers, I could
not contemplate the delicate conformation of the
roots, leaves, and fruit without admiration. Can that
Being (thought I) who planted, watered, and brought
to perfection, in this obscure part of the world, a
thing which appears of so small importance, look
with unconcern upon the situation and sufferings of
creatures formed after His own image? Surely not.
Reflections like these would not allow me to despair.

I started up, and, disregarding both hunger and fatigue, travelled forwards, assured that relief was at hand—and I was not disappointed." The special moss on this occasion, although we know not upon what evidence, is stated by Dr. Johnson to have been *Dicranum bryoides*.

NOTABLE MOSSES.

Young Mr. William Hooker was, like many other young men dwelling in the country, very fond of birds, and of studying their habits, so that he became, having plenty of leisure, a good practical ornithologist, but not much interested in botany. We believe that he was living at this time in or near Norwich. During one of his country walks over Mousehold Heath, and through Sprowston on the road towards Rackheath, he entered a little wood, and sat down to listen to the notes of some bird. Whilst thus occupied, his attention was riveted by a little moss at his feet, which seemed to him the strangest little plant he had ever seen. It was almost, if not entirely, destitute of leaves, or at least with only a small rosette of minute ciliated leaves, but a comparatively large and conspicuous reddish capsule. He collected some specimens, determined to ascertain its name and history, which resulted in the discovery that it was called *Buxbaumia aphylla*, and that it had never previously been found in England. This little incident led Mr. Hooker, afterwards Sir William Hooker, and the director of Kew Gardens, to turn his attention to botany, and especially to the Higher Cryptogamia, amongst which he laboured earnestly and long.

What a small event may alter the whole current of a life, and make or mar its future !

We are unable to furnish any extensive catalogue of notable mosses, and their historical associations, but there may be a few which are notable in other ways, and especially for properties which they possess, or are credited with possessing. We remember none which have even found a place in domestic medicine, or had the reputation of effecting a marvellous cure. Only one indigenous little species has been invested with the mystery of shining in the dark.

The pretty little moss *Schistostega osmundacea* is often quoted as exhibiting exceptional properties of refraction, and some have said of phosphorescence. It loves the shade of caverns, " which are sometimes lighted by a golden-green gleam from the refraction of the confervoid shoots." This is, perhaps, the most careful way of stating the phenomenon. Others say that " the young plant, when growing in caves, emits a beautiful golden-green light." It appears that the confervoid prothallus was formerly described by Bridel as an alga, and it was long supposed to be phosphorescent, but this error was at length dispelled, and the luminous appearance is now believed to arise from " the condensation and reflection of the little daylight admitted, by the pellucid convex cellules of the prothallium."

Other mosses have been specially alluded to for possessing unusual hygroscopic propensities. This is observed more especially in the beautiful fringe of teeth which surround the mouth of the capsule. So

delicate and susceptible are they in some cases, as to
be influenced by the moisture of the human breath.
They may be closely adpressed, or twisted at one
instant, so as to conceal the opening of a capsule;
in the next instant they become sensitive to the
moisture which falls upon them, gradually untwisting
and falling back, like the tentacles of a miniature
sea-anemone, and then, as the moisture evaporates,
closing up again, as they were before—

> " When Heaven's blithe winds had unfolded them,
> As mine-lamps enkindle a hidden gem,
> Shone smiling to Heaven, and every one
> Shared joy in the light of the gentle sun."

LIVERWORTS.

THIS is a small order of cryptogamous plants, related to the mosses, and sometimes vaguely, but inaccurately, included with them. In appearance, the typical forms at least, resemble mosses, in having organs analogous to a stem and leaves ; but differing in the more delicate and filmy substance of those leaflets. More characteristic still is the difference in fruit, which in mosses is mostly contained in a capsule, opening by a lid, with a fringed mouth, and in liverworts with a splitting receptacle (Fig. 9) but without a lid. These distinctions, however, need not be insisted upon here, because short general descriptions are never satisfactory, and this is a technical question, outside our present limits. It may be added, however, that the spores within the receptacle in liverworts, are often mixed with spiral thread-like organs, called *elaters*, which are absent in mosses.

If we go back to Lindley's " Vegetable Kingdom " of 1847, we ascertain that the estimated total of species of the *Hepaticæ* at that period was seven hundred. The present number can only be guessed, but, on the basis of certain knowledge, it may be assumed at not less than two thousand, although we believe it to be more.

Some of the species have a form resembling the
thallus of certain lichens, but of a
different colour; others, again, possess
a stem and leaves, but the latter
often assume strange outlines, accompanied by curious appendages and
stipules; they are also more diapha-
nous than is usual with mosses, and
in colour more frequently assume a
purplish hue. Practically, with the
slightest experience, it is not difficult
to discriminate between them.

The capsules, or spore-fruits in the
liverworts, are preceded by the production of certain organs known as
archegonia and the *antheridia:* from
the former of these the capsule has
its origin; while the antheridia, after
discharging their moving spiral filaments, or the spermatozoids, die away
and disappear. In the frondose liverworts, such as *Riccia*, the archegonia
and antheridia are developed in the
cellular substance of the frond; in

Fig. 9.—Capsule of
Jungermannia.

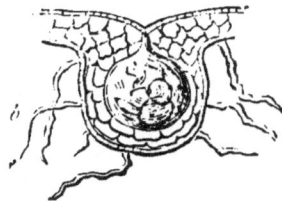

the leafy *Jungermanniæ* they are found in the axils

Fig. 10.—*Riccia fluitans.* *a*, nat. size; *b*, imbedded spore case magnified.

of the leaves, or at the apex of the stem, but under all circumstances of a common type. The antheridia present themselves at first as cellular papillæ, which grow out and become converted into stalked club-shaped, or at length almost globose bodies. At a certain epoch this stalked sac becomes ruptured above, and the contained cellules emerge. In the interior of each of these is seen a spiral filament coiled up, which soon breaks out of the cellule and exhibits active rotatory motion. These spiral filaments are the so-called spermatozoids.

The archegonia are more variable. In *Pellia* they grow out from the under side of the edge of the frond, as a flask-shaped body, with an enclosed cellule which, after impregnation, becomes developed into the spore-bearing capsule. In the leafy species they are similar flask-shaped bodies. In *Riccia* both antheridia and archegonia originate very early, and the parenchyma of the frond grows up around them as they advance in development. Hofmeister states that he has several times seen spermatozoids swimming about the archegonia of species of *Jungermannia*, when brought quickly under the microscope. He also found at the mouths of archegonia, in other species, more or less curled colourless filaments which resembled spermatozoids, but were motionless.

ELATERS.

Lindley, writing of the Hepaticæ, observes that " a remarkable point of structure in the liverworts is the spiral filament, or elater, as it is called, lying among the sporules within the spore-case. This consists of a single fibre, or of two twisted spirally in different directions, so as to cross each other, and contained within a very delicate, transparent, perishable tube; they have a strong elastic force, and have been supposed to be destined to aid in the dispersion of the sporules—a most inadequate end for so curious and unusual an apparatus. It is more probable that they are destined to fulfil, in the economy of these plants, some function of which we have no knowledge." The elaters which occur in the Horsetails, or *Equisetaceæ*, are of a different character to those found in the liverworts, and should not be confounded with them. In this case the elliptical spores are furnished with four elastic filaments, attached about the middle of one side, which are coiled once or twice round the spore, before it is discharged from the capsule, in the position where they were originally developed; but when the spore is discharged they uncoil with elasticity, causing the spore to be jerked away. The result of investigation appears to prove that the outer coat of the spore splits by a spiral fissure, and, separating in narrow ribbons from the inner coat, becomes the elastic appendages to which the name of "elater" has been applied.

In the liverworts the elaters are more or less elongated membranous tubes, which are closed at the ends, and contained within them are one or more elastic spiral fibres. These bodies occur mixed with the spores within the capsules, and are sometimes attached by one end to the valves of the capsule, so that they not uncommonly remain behind after the spores are dispersed. No such structures are met with in mosses.

Griffiths says that they are by no means universal in the Hepatics, and hence too much importance has, in his opinion, been attached to them. "They are almost universally associated with the existence of fibrous cells of the capsule ; and this would seem to corroborate the truth of Mirbel's conjecture, that they are modifications of such cells. In fact, the transition between these cells and the elaters is very evident in *Pellia epiphylla.* In their younger states they consist of an elongated cell, containing one or two grumous nuclei. They are not to be confounded, as has been done, with the remains of cellular tissue, which occurs intermixed with the sporules in some genera. Too much stress, he thinks, has been laid on these organs as inducing the dispersion of the sporules, neither did he see, in any instance, that the sporules adhere to the elaters." And, again, when writing of *Marchantia,* he says, "Unless the elaters germinate, I cannot imagine any special use they may be of, because no means are visible to further their association with the sporules—the only organs for which they can with reason be supposed to be provided. They are subject to precisely the same contingencies as the sporules, at the dehiscence of the fruit, a period when they are in their state of greatest perfection."

LICHENS.

THE lichens are manifestly closely related to the fungi in many particulars, especially of structure, and were so regarded by Berkeley, who classed Fungi and Lichens together as two great families constituting the Mycetal Alliance. They differ from fungi in mode of growth, in their living at the expense of the surrounding medium, and not of the matrix, and in the production of gonidia. The greater proportion of fungi are destructive, either of living structure, or dead matter by disintegration, whereas lichens are not destructive of the substances upon which they flourish, or rather to which they are attached. Moreover, lichens are usually of slow growth, persistent for years, not being so prone to decay.

It is difficult, in the absence of any complete catalogue of recent date, to estimate with any precision the number of described species ; but we know them to be far less numerous than fungi or algæ. In 1847 the estimated number of species for the whole world, according to Lindley's "Vegetable Kingdom," was 2400. In 1869 Krempelhuber's catalogue raised the number to 5131, and since that period a great many new species have been enumerated, so that we may fairly consider that, during nearly a quarter of a century, the number has been raised to at least eight thousand species, or probably about equal to that of the mosses. Very few of the lichens are of any economic value, since the advance in chemistry has superseded the dye-lichens, to a very great extent, by artificial products. As articles of food they were always miserable substitutes, and in medicine more fanciful than of real value.

One important feature may be noted in which lichens differ greatly from other cryptogamic forms, and it is a difference of great importance to remember. It is thus stated by Dr. Lauder Lindsay : "Lichens are perennial ; they grow very slowly, but they attain an extreme age. Some species growing on the primitive rocks of the highest mountain ranges in the world, are estimated to have attained an age of at least a thousand years ; and one author mentions, after the lapse of half a century, having observed the same specimen of *Sticta pulmonaria* on the same spot of the same tree."

Efforts have been made, on more than one occasion, to demonstrate that the growth of lichens upon trees is injurious to the tree. This appears to be controverted by one small fact. Pharmacists are aware that the most valuable kinds of cinchona bark, as for instance " Crown bark," are habitually covered with lichens, and that this covering favours the development of the alkaloids. The inferior kinds or bark are not inhabited by lichens. Acting upon this suggestion, a practice has arisen in cinchona plantations of covering the bark with moss, as a substitute for lichens, and the bark thus mossed has been found to produce a larger percentage of alkaloids than unmossed trees.

Lichens are not popular objects, and their study is almost absolutely confined to a few scientific men, yet there are many reasons why a knowledge of them should be more universally diffused. One writer remarks that "if we consider that many species have a texture which, by readily imbibing and eagerly retaining moisture, renders them in a sense independent of all climatal changes, enabling them equally to brave polar cold and tropical heat; that many not only cling with such tenacity as to be inseparable from, but can corrode or disintegrate the hardest and barest rocks, even pure quartz; that the most ample provision has been made by the great Author of all for their reproduction or multiplication, in spite of the most adverse external circumstances, and under conditions fatal to all higher vegetation, both by the multiplicity and abundance of their reproductive cells —which sometimes constitute almost the entire bulk of the plant,—the extremely minute size and delicate nature of those cells, by virtue whereof they are disseminated by every shower or zephyr, and the readiness with which these germinate; we cannot fail to increase our surprise that a curiosity has not been sooner awakened to become familiar with the natural history of plants which strew the path of man wherever he roams over the wide world—which constitute the most universally diffused type of terrestrial vegetation." To this one answer may be given, as regards British teachers and British students, that unfortunately, hitherto, no successful effort has been made for its popularization ; almost all, if not all, who have written upon the subject have never really sympathized with the student, or appreciated his wants ; and hence British Lichenology has been crude and rigidly technical, to a degree which has never been attained in any other branch of botanical science.

In the coldest as well as the hottest regions yet visited, and at the greatest heights yet reached by man, lichens have been met with in more or less abundance. There seems to be a great similarity of species in different parts of the world. Two-thirds of New Holland lichens are natives of Europe, and the majority of Himalayan species are European forms.

On the highest mountains, between an elevation of thirteen thousand and sixteen thousand feet, there is a terminal region of vegetation known as the zone of lichens. On the central and southern Alps, above the highest limit of flowering plants, species of *Parmelia*, *Lecidea*, etc., are found on all the rocks projecting through the snow, and they occur at above sixteen thousand feet on Chimborazo. One species, called the Geographic Lichen, is found above the line of perpetual snow on the Alps, and is the last type of vegetation on the Andes and Himalayas ; and Dr. Hooker found lichens as the last remnants of vegetation in the southern hemisphere. Although found at such extreme limits of temperature, the number of species are limited, and in great contrast to their development in the tropics. In Spitzbergen, at about thirty degrees latitude, only thirty species were found, whilst at Madagascar, under the tropic of Capricorn, they numbered five thousand. Hence it may be inferred that these cryptogams are the most widely diffused of all terrestrial plants.

DYE LICHENS.

The most important economic use to which species of lichens have been applied in the past was that of the production of certain purple dyes, much in vogue before the introduction of aniline. Dr. Lauder Lindsay contended that these dyes were known from a remote antiquity, and that they were alluded to in Ezekiel (xxvii. 7), "Blue and purple from the isles of Elishah was that which covered thee." The general name of these dyes was orchil, or orchella; and orchella weed was long an important item in our trade imports—reaching in value to from £60,000 to £80,000 per year. The various species of *Roccella* were imported, under the names of Angola, Lima, Cape, or Canary orchella weeds, for the manufacture of orchil or cudbear. The English name was *orchil*, the Scotch *cudbear*, and the Dutch *litmus*. The first was manufactured in the form of a liquid or paste, of a rich purple colour; the second occurred in the form of a powder, of a crimson tint; whilst the third was only known in the form of small oblong cakes of an indigo-blue colour. The colour was naturally reddish, the blue tint being communicated by the addition of alkalies. The principle of the manufacture consisted in reducing the cleansed and powdered lichen into a pulp with water, adding thereto an ammoniacal liquid, chiefly gas-liquor, although another kind of ammoniated liquid was largely collected and employed, even within our own experience. The mass was macerated in a moderately warm place for various periods, from a few days to several weeks. By this means a kind of fermentation was induced, and at the end of the process a beautiful purplish compound was produced, possessing a peculiar ammoniacal odour.

A testing process may be adopted in a small way by macerating fragments of lichen in a vial containing liquid ammonia, or common hartshorn and water, and shaking them together.   If any of the colouring matter be present, the liquor will soon acquire a reddish tint, which, by standing longer, will become of a rich purple.

In some parts of Scotland there still remain, not only the tradition, but the practice of domestic dying with the "crottles."   Formerly a cudbear manufactory flourished at Leith and Glasgow, which is now extinct, and then large quantities of *Lecanora tartarea* were collected by the peasantry of the western highlands and islands.   Dr. Lindsay says that "in Scotland, not many years ago, particularly in certain districts, almost every farm and cotterhouse had its tank or barrel of 'graith,' or putrid urine (which was the form of ammoniacal liquor employed), and its 'lit-pig,' wherein the mistress of the household macerated some familiar 'crottle' (the Scotch vernacular for the dye-lichens in general), and prepared therefrom a reddish or purplish dye."

In addition to these purplish colorific substances,
some of the lichens possess yellow or greenish colouring matter, also of an acid nature, such as those due
to vulpinic acid, produced in the thallus of *Cornicu-
laria vulpina*, and the parietinic acid found in the
common wall lichen, *Parmelia parietina*. Brown
colouring matter is also abundantly present in many
lichens ; and the domestic woollen hosiery was com-
'monly dyed at home in the Highlands with some of
the crottles, although they were never sufficiently
valuable to be applied commercially on a large scale.
One of these was the " Oak-lungs," *Sticta pulmonaria*,
found growing on oak trees (Fig. 11).

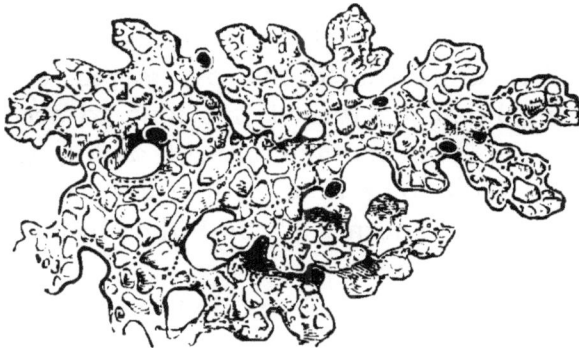

FIG. 11.—Oak-lungs.

The species of *Roccella* which constitute the true
orchella weed, grows chiefly on maritime rocks, but
in some foreign countries it is found on trees, notably
on the mango tree in parts of India. It is found
very sparingly on the south coast of England, in the
Channel Islands, and on the adjacent islands and
coasts of France. In tropical Africa, Asia, and
America it reaches its highest development, and on

the coasts frequently attain great size, and a
tough leathery consistence. The most valuable,
commercially, was formerly the Angola weed, from
the Portuguese settlement of Angola, in South
Africa. While Italy enjoyed a monopoly in the
manufacture of orchil, large quantities were supplied
by Teneriffe, the Canaries, Azores, and neighbouring
islands ; the inhabitants farmed out the right to
gather the orchella weeds, paying therefor considerable sums to the Government. Prior to this
these lichens were only known in the islands, and on
the shores of the Levant. It is affirmed that the
capability of their yielding a dye, by maceration in
ammonia, was discovered accidentally by a Florentine
merchant who was travelling there, and who observed
that putrid urine tinged the plants red or purple.
Returning home he turned this hint to account by
commencing the manufacture of orchil, which he
carried on for a long time with great secrecy, under
the name of "Tournesol," and thereby acquired a
considerable fortune.

This little episode will exhibit how a very small
and comparatively insignificant plant, which for a
long time appeared to have no special mission upon
earth, and seemed to be of no value to man or beast,
became the nucleus of an important industry, and,
for very many years, supported a host of collectors,
and employed a considerable population, in the
manufacture and application of orchil dyes.

REINDEER MOSS.

Amongst what are termed the lower orders of the

vegetable kingdom there are few species which can compete for interest and utility with the lichen known as the Reindeer Moss, *Cladonia rangiferina.* It varies much in size, being seldom taller than three or four inches in Britain, but two or three times as long in more northern climes. It is almost a cosmopolite, but its geographical range in various parts of the world is very irregular and limited. In Lapland it covers vast tracts of country, growing to a height of from six to twelve inches. The barren plains so covered are the favourite and only pastures of the reindeer during winter. The animals clear away the snow by means of their horns, to browse on the lichen. On the destruction of the forests by fire this plant continues to grow, and then reaches its greatest luxuriance. It would be impossible for the reindeer to exist in these climates during the winter were it not for this plant. The Laplanders are in the habit of collecting it with rakes in the rainy season, when it is flexible and readily separates from the ground where it has grown. It is then laid up in heaps to serve as fodder for the cows. Parry, in the "Narrative of his Fourth Voyage," mentions his officers collecting supplies of this lichen as provender for the reindeer, which he employed in the capacity of horses. He says that it required a great deal of picking to separate it from the moss with which it usually grows. The daily quantity of cleaned Reindeer Moss necessary for each animal on a journey, he estimates at four pounds ; but, he remarks, it can easily remain for five or six days without food. To prepare it as fodder for cattle, in some northern countries, hot water is poured over it; it is then mixed with straw, and a little salt sprinkled over the mixture. Cattle so fed are said to produce delicious milk and butter, while their flesh becomes fat and sweet. The stag, deer, roebuck, and other wild animals also feed on it abundantly during winter.

Dr. Clarke and his companions, during his travels in Lapland, were tempted to eat some of this lichen. "To our surprise," he says, "we found that we might eat of it with as much ease as of the heart of a fine lettuce. It tasted like sweet bran. But after swallowing it there remained in the throat, and upon the palate, a gentle heat or sense of burning, as if a small quantity of pepper had been mixed with the lichen. We had no doubt that if we could have procured oil and vinegar it would have made a grateful meal. Cooling and juicy as it was to the palate, it nevertheless warmed the stomach when swallowed, and cannot fail of proving a gratifying article of food to man or beast during the dry winter of the frigid zone. Yet neither Laplanders nor Swedes eat of this lichen." It is elsewhere stated, but not by what nationality, that it is sometimes powdered, mixed with flour, and baked into bread, or it is boiled in milk or broth.

On the authority of Dr. Clarke, it is related that "when Gustavus III. succeeded to the throne, an edict was published, and sent all over Sweden, recommending the use of this lichen to the peasants in time of dearth, and they were advised to boil it in milk." The starchy component of this, and such similar lichens as the Iceland Moss, is *lichenin*, or lichen starch, which is said to be intermediate in composition and character between dextrine and common starch. Dilute and boiling sulphuric acid converts it into sugar, while nitric acid transforms it into oxalic and saccharic acids.

LICHEN MANNA.

It is generally considered that the Israelites in the desert must have had previous knowledge of some substance under the name of manna, and therefore applied that name to the miraculous supply recorded in Exodus. Be that as it may, substances having different origins have long been known in the East under the general name, and there has not been wanting even the assertion that one or other of these was the genuine " manna of the desert." To one only of these substances does the scope of the present volume permit us to allude, and that is a peculiar form of lichen which has, from time to time, been reported to descend like a shower of rain. " Several occurrences of what is called a fall of *manna* are attributable to the accumulation of this lichen, *Lecanora esculenta.* Aucher-eloi observed it in Persia, in layers of nearly four inches in thickness. He sent specimens, with the following note to France : ' In 1829, during the war between Persia and Russia, there was a great famine in Oroomiah, south-west of the Caspian. One day, during a violent wind, the surface of the country was covered by a lichen, which *fell from heaven.* The sheep immediately attacked and devoured it eagerly, which suggested to the inhabitants the idea of reducing it to flour, and

F

making bread of it, which was found to be good and
nourishing.   The country people affirm that they
had never seen this lichen before nor after that time.
During the siege of Herat (which is about 876 feet
above the sea), more recently the papers mentioned a hail of *manna* which fell upon the city,
and served as food to the inhabitants.   A rain of
manna occurred April, 1846, in the district of Jeni-
schehir—the government of Wilna—on the grounds
of M. Tizenhaux, and formed a layer three or four
inches in thickness.   It was of a greyish white colour,
rather hard, and irregular in form, inodorous and
insipid.

"Pallas observed it in the mountainous, arid, and
calcareous portions of the great desert of Tartary.
M. Eversham collected it in the steppe of the Kirghiz
to the north of the Caspian Sea, where it is called
semljenoichleb.   M. Ledebour has observed it in the
same countries, but chiefly those which border on
Altai ; and Bilezikdgi saw it also in Anatolia, in
1845.   Dr. Léveillé gathered it in the Crimea, and
Dr. Guyon recently in Algeria." [1]

A more recent and detailed account was forwarded
from Erzeroum, in 1849,[2] which supplies some very
interesting particulars.   It states that "two months
previously a report was current in Erzeroum that a
miraculous fall of an edible substance had occurred
near Byazid, but as the simplest facts are often
greatly distorted and exaggerated in this country,

---

[1] "Notices of Koordistan," by A. H. Wright, in *Silliman's Journal*,
2nd series, vol. iii., May, 1847.

[2] *Gardener's Chronicle*, Sept. 15, 1849, p. 581.

and the most unblushing falsehoods circulated, in connection with anything of unusual occurrence, the
European residents here were not inclined to listen
credulously to the accounts of this 'wonderful fall
of bread from heaven.' The report, however, instead
of being soon forgotten, gained daily more ground,
specimens of the substance were brought hither, and
travellers from Byazid bore testimony to the fact of
several showers of these lichens having taken place.
Finding that there was some foundation for this
phenomenon, I thought that the matter was deserving
of investigation. I therefore applied to Dr. Heinig,
the sanitary physician at Byazid (the only European
residing there), to furnish me with information, which
I elicited by means of a series of questions, of which
the following is the result :—

"About the 18th or 20th of April last, at a period
when there had been, for a whole fortnight, very
rainy weather, with strong winds from the S.E. and
E.S.E., the attention of the shepherds and villagers
frequenting the country near Byazid was attracted
by the sudden appearance, in several localities, of a
species of lichen, scattered in considerable quantities
over certain tracts, measuring from five to ten miles
each in circumference. Dr. Heinig describes two of
these spots as follows : One is situated three miles
east of Byazid, behind a range of rocky mountains,
stretching from the north, gradually towards the
south-east. The other is five miles to the south of
Byazid, near a similar range of rocks, running in the
above-named direction.

"It is remarkable that no one had before observed

these lichens in the neighbourhood, not even the shepherds, who often pasture their flocks on the crags, and in almost inaccessible places; and Dr. Heinig, who has been on Mount Ararat (which is close to Byazid), and who appears to have a taste for rambling over mountains, says he has never met with any. What seems to confirm the assertion that these products were not known previous to their unaccountable appearance is, that last year the crops were greatly injured by locusts, and a famine threatened; and had the substance been known to exist anywhere in the vicinity, it would most assuredly have been eagerly sought after, and collected last autumn, when the price of wheat had risen to more than double its usual value. A similar phenomenon is said to have occurred at Byazid some years ago, when it is probable that the edible quality of these lichens became known to the natives; unless showers took place previous to that period, which I have not been able to ascertain.

"No proof has been adduced of any one having seen the substance fall, but as the first intelligence was brought by villagers who, early one morning, had observed the lichens strewed over a tract of ground where they had not observed any on the evening before, it is probable that the showers must have taken place during the night. In some localities, the one or the other kind of lichen alone was found, in others the two species mixed. On the 19th of June another quantity of lichen was discovered, and as the spot was a well-frequented one, it seems likely that the fall had occurred only a few days previously.

"From all accounts the quantities collected have been very great. Dr. Heinig says that a person could collect at the rate of a pound and a half in an hour, which, considering the lightness of the product, is a tolerable quantity. The substance is ground up with wheat, and made into bread, or eaten simply in its raw and natural state."

Commenting subsequently on the above narrative in the journal in which it appeared, the Rev. M. J. Berkeley[1] referred to previous occurrences of like phenomena, some of which are alluded to above. He says that Treviranus, in 1815, figured specimens obtained by Blume from somewhere to the east of the Caspian Sea. Parrot brought some which were collected in the beginning of 1828, and said to have descended from the skies in some districts of Persia and to have covered the ground to the depth of five or six inches. Göbel analyzed some of these, and believed them to have been carried by electric winds from distant localities. Ledebour met with the production frequently in the Kirghiz steppes and in Central Asia, and had seen them spring with great rapidity after repeated heavy rains, and believed that they must have been produced under similar circumstances in Persia, and it is observable that in all the accounts the supposed descent is uniformly during rainy weather.

Eversmann, who had an opportunity of studying the species on the rivers Emba and Jaik, and also near Lake Aral, was convinced that, even in the earliest stage of growth, there is not the slightest attachment

[1] *Gardener's Chronicle*, Sept. 29, 1849. p. 612.

even to a grain of sand, but that the thallus is developed freely, as was at first declared by Pallas.
Treviranus says that specimens supposed to have
descended from the clouds at Mount Ararat exist
in the Museum of the Armenian convent of St.
Lazzaro, in an island of that name near Venice.

This curious production is eaten both by men
and animals in the several countries extending from
Algiers to Tartary, where it is produced. The sheep,
however, which feed upon it in Algiers, it is said, do
not thrive, in consequence, it is supposed, of the large
amount of oxalate of lime which it contains, amounting, according to Göbel's analysis, to nearly sixty-six
per cent. The individual plants weigh from a few grains
to two scruples or
upwards, even when
dry, and when
swollen with mois-
ture nearly twice
as much. Of the
two species *Leca-
nora esculenta* is
the one which was
known to Pallas,
and which has the

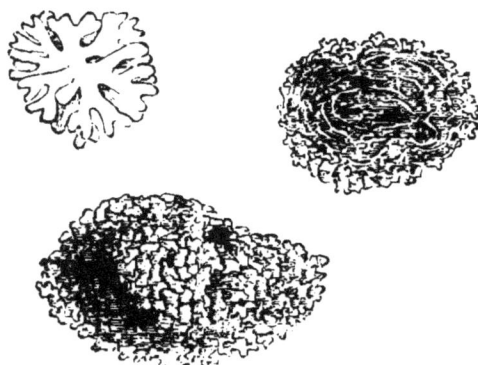

Fig. 12.—*Lecanora affinis.*

central portion consisting of loose threads, which
gradually become more densely packed towards the
circumference, forming an extremely close cellular
network. In *Lecanora affinis*, which is the second
form received from Erzeroum, the whole substance is
compact, with few, if any, free threads.

Tripe de Roche.

It is well known that in Northern countries, and especially in the arctic regions, a large, coarse, and leathery lichen is produced on the rocks, which is familiar to the Canadians as *Tripe de Roche*, or Rock Tripe. Botanically the species are known as *Gyrophora*, mostly of a grey colour, almost black, expanded and leaf-like, with a tubercular surface not much unlike leather. To appearance they are uninviting as articles of food, and yet under dire calamities they have been useful in supporting human life. The service they rendered to Sir John Franklin's party have invested them with something of romance, which otherwise they could scarcely have deserved. There is something pathetic in the references to this abominable food in the journals of the expedition. "Near Hood's River the surface of the large stones were covered with lichens of the genus *Gyrophora*, which the Canadians term *Tripe de Roche*. A considerable quantity was gathered, and, with half a partridge each, furnished us with a slender supper."[1] "The Tripe de Roche, even where we got enough, only served to allay the pangs of hunger for a short time."[2] "This unpalatable weed was now quite nauseous to the whole party, and in several it produced bowel complaints."[3] "The Tripe de Roche had hitherto afforded us our chief support, and we naturally felt great uneasiness at the prospect of being deprived of it, by its being so frozen as to render it impossible for us to gather

[1] "Journals," p. 403. [2] Ibid., 407. [3] Ibid., p. 408.

it."[1] And on one occasion it is remarked that the party had no other food for eleven days. Dr. Lindsay states that species of *Umbilicaria* or *Gyrophora* are frequently eaten in periods of scarcity in Iceland, as a supplement to the more nutritious "Iceland Moss." He adds that their nutritive properties depend on the presence of a large amount of starchy matter, and that, when boiled, they yield a firm nutrient jelly, accompanied by a bitter principle which is possessed of purgative properties.

Reference has been made quite recently to an edible lichen not far removed from those which furnish the Tripe de Roche, but we fear that the glowing eulogy is an exaggeration. The statement is as follows: "*Endocarpon miniatum*, which is found in many of the United States, is also found in Japan and Cuba, two widely diverse localities. It inhabits calcareous rocks, and may easily be mistaken for *Umbilicaria*, two species of which it resembles. Mr. Minakata, who is a distinguished scholar and naturalist, and who has lately spent two years in the United States in study and travel, has informed the writer that large quantities are collected in the mountains of Japan for culinary purposes, and largely exported to China as an article of luxury. He expressed surprise that no attention was paid to it in America. The name by which it is known in Japan is *iwataka*, meaning 'stone mushroom.' Properly treated it resembles tripe."[2]

This latter lichen is also found in Britain, in low-

[1] "Journals," p. 445.
[2] Prof. W. W. Calkins, in *Botanical Gazette* (1892), vol. xvii. p. 418.

land and sub-alpine situations, on moist rocks in the
neighbourhood of waterfalls or rivers, or is frequently
covered by water. Dr. Lindsay found it by the side
of the Tay, on boulders frequently covered by the
river when flooded, and on the craggy southern face
of Kinnoul Hill near Perth. When under water it
has a deep olive colour. It sometimes attains a
diameter of several inches

DUALISM.

We cannot exclude hypotheses from a general pur-
view of the "curiosities," if not the "romance" of
lower vegetable life; indeed, two or three instances
are present to our minds in which hypotheses, popular
for a time, because novel or strange, might be regarded as unqualified romance, which a future generation will contemplate as fairy tales. One hypothesis
of this kind created some excitement twenty-five
years ago, but has at length almost subsided into
oblivion. Our record of this hallucination will be
brief. The high priest Schwendener thus expressed
his dream : "As the result of my researches, all these
growths (lichens) are not simple plants, not individuals, in the ordinary sense of the word ; they are
rather colonies, which consist of hundreds and thousands of individuals, of which, however, one alone
plays the master, whilst the rest in perpetual captivity
prepare the nutriment for themselves and their master.
This master is a fungus of the class *Ascomycetes*, a
parasite which is accustomed to live upon others'
work ; its slaves are green algæ, which it has sought
out, or indeed caught hold of, and compelled into

its service. It surrounds them as a spider its prey, with a fibrous net of narrow meshes, which is gradually converted into an impenetrable covering; but whilst the spider sucks its prey and leaves it dead, the fungus incites the algæ found in its net to more rapid activity, nay, to more vigorous increase." This may be all very poetical, but it is not very explicit, and needs a commentary. This we furnished very many years since, and it has never been called in question, but corroborated, to the following effect: that the two great points sought to be established are these, that what we call lichens are compound organisms, not simple, independent vegetable entities; and that this compound organism consists of unicellular algæ, with a fungus parasitic upon them. The coloured gonidia which are found in the substance, or thallus, of lichens, are the supposed algæ, and the cellular structure which surrounds, encloses, and imprisons the gonidia, is the parasitic fungus, which is parasitic on something infinitely smaller than itself, and which is entirely and absolutely isolated from all external influences. In plain words, the gonidia are algæ, and all the rest of the plant is fungus.

This hypothesis has so few adherents, and those few of such little eminence or authority, that it would be quite unnecessary and impolitic to reopen the discussion or repeat the arguments against it, which have never been successfully controverted. As a mere historical summary, it may be advisable to state a little more definitely what were the points in dispute, and, in doing so, to intimate, as a matter of fact,

that no lichenologist of any repute, and no mycolo-
gist of any authority, supported the theorists during
the controversy, but rejected their views as untenable.
This is important, as an historical fact, since these
were just the scientific men who were practically the
best acquainted with the organisms in question, and
whose judgment would have the greatest weight.
Lichens consist, normally, of a thallus, or vegetative
system, which in many species is a tough coriaceous
expansion, horizontal or vertical, attached at the base
for its support to rocks, stones, wood, and other substances, but *not* deriving sustenance from the object
to which it is attached. In the interior of this thallus
minute green bodies, or cellules, form a sub-cortical
layer, which are denominated *gonidia*, and which Dr.
Nylander, a most eminent lichenologist, says, " constitute a normal organic system, necessary and of the
greatest physiological importance, so that around
them we behold the growing (or vegetative) life
chiefly promoted and active." In addition to these
is the reproductive system, which consists of apothecia, or discs, borne upon the thallus, containing the
asci and sporidia, or reproductive organs. Such are
the organisms which have been called lichens, and
are regarded by the best scientific authorities as complete and autonomous plants, equally with fungi and
algæ, amongst the lower cryptogamia. On the other
hand, the theorists contended that the thallus and
reproductive system were, not only fungoid in character, but absolute fungi, whilst the green gonidic
layer of cellules were simply unicellular algæ, *upon*
which the fungus was parasitic. Hence the lichens

were not autonomous plants, but mere compounds of fungus and algæ.

This contention was opposed on the following grounds: and, firstly, that the so-called fungoid element was not a fungus, because lichens are perennial; they grow very slowly, but they attain an extreme age. Some species, growing on primitive rocks of the highest mountains in the world, are estimated to have attained an age of at least a thousand years; and one author mentions, after the lapse of half a century, having observed the same specimen of *Sticta* on the same spot of the same tree. On the other hand, fungi are mostly annual, very short-lived, their whole existence being limited to a few weeks, rapid in growth, and rapid in decay, not a trace of some species remaining after a few days.

Lichens will exist under conditions of aridity which no other vegetables could support. Some are peculiar to calcareous rocks, a few on arenaceous rocks, many are common on the granitoid series, and one, *Lecidea*, is frequent on the purest and smoothest quartz. Fungi, on the contrary, must have moisture for their very existence' sake, are mostly found in damp and shaded situations, and could never exist under the conditions above enumerated for lichens. Of all plants lichens support extreme cold most successfully, whilst fungi succumb at the approach of frost.

Lichens which grow upon the bark of trees may be seen flourishing in profusion during the life and vigour of the tree. Cinchona bark which has been covered with lichens during growth has its qualities improved, whereas the portions attacked by fungi are valueless from the tissue being destroyed by the mycelium. Fungi do not, and cannot, flourish on growing and vigorous bark, but on diseased, dead, or decaying spots.

Lichens obtain the greater portion of their pabulum from the atmosphere, and only their mineral constituents from the matrix. Fungi obtain their chief support from the decaying vegetable matter on which they flourish, gathering up the nitrogenous results of decay, and disintegrating the matrix on which they prey. Lichens are preservers, whilst fungi are destroyers.

Lichens, in their chemical composition, contain a large number of substances which are wholly unknown amongst fungi. Likewise they hold but a small percentage of water, as compared with fungi, so that in dessication they do not shrivel, collapse, or diminish perceptibly in size ; whereas fungi shrivel, and collapse, and are constantly liable to the attacks of insects, or, if damp, to the development of mould. Lichens may be preserved for years without fear of insect or mould.

The contention was also opposed on the ground that the gonidia are a part of the lichen structure, and are consequently not appropriated green algæ. Lichenologists, with one consent, admit them as essential parts of lichen structure. One says that they may be regarded as intermediate in function between the vegetative and reproductive cell, assuming the offices, and partaking of the characters of both. Nylander contends that the absurdity of the hypothesis is evident from the very consideration that it

cannot be the case that an organ (gonidia) should at
the same time be a parasite on the body of which it
exercises vital functions. If the gonidia are true
algæ, it was contended by the opponents of the
hypothesis, it is insufficient to state that they so
closely resemble algæ that they might be mistaken
for such ; there must be some undoubted evidence
produced that they are algæ in fact, and not in
appearance. It has been demonstrated that the
gonidia are developed within the substance of the
lichen itself, in a determinate and uniform manner ;
that, instead of being altogether a foreign substance,
they are generated within it, and hence, according to
the theory, the parasite produces from its own substance the host upon which it is parasitic, which is
an absurdity. "Parasitic existence is autonomous,
living upon a foreign body, of which nature prohibits
it from being at the same time an organ."

From the foregoing and similar facts, it was concluded that the assumption that two separate and
distinct organisms are combined in one plant, which
by its own proper system of reproduction is capable
of continuing the species, each individual of its
progeny also exhibiting the same phenomena of dual
existence, is inconsistent with known scientific facts,
because whilst one supposed plant proceeds from its
proper germ, the other has none, and is, therefore,
spontaneously evolved.

The theory assumes further, that a fungus is parasitic upon a smaller and weaker organism, which it
does not injure or destroy. This again is contrary
to scientific fact, as it confers upon, or assumes for
a destructive force conservative powers.

It was assumed that the green gonidia were uni-cellular algæ, but the assumption was not proved or their independent existence demonstrated ; but it has been shown, on the contrary, that they are a component part of the lichen-thallus, being definite organs of the lichen, performing definite functions in its behalf. Hence it was contended that where an organism is seen to exist, and continues to reproduce itself in conformity with a certain type, it is folly to attempt, by multiplying causes, to account for phenomena in an abnormal manner which can be readily accounted for by normal causes. When we see an organism in possession of certain organs, which organs perform certain functions however small, and these organs and functions are inherited and transmitted, there is a pretty strong presumption that all our efforts to demonstrate external causes, to account for phenomena already well accounted for, will only embark us on a profitless speculation.[1]

[1] For further information on this vexed question, consult " The Dual-Lichen Hypothesis," by M. C. Cooke (*Grevillea*, March, 1879), p. 102 ; *Journal of Quekett Microscopical Club* (1879), vol. v. 170.

ALGÆ.

UNDOUBTEDLY the Algæ form a very large
section of the lower cryptogamia, although at
present it is difficult to arrive at any definite census
of their number.   The exceptions are so few that
they may be characterized as truly aquatic, inhabit-
ing either fresh water or salt, throughout the world ;
but, although often attached by a rooting base, yet
destitute of true root, and not obtaining any sustenance from the object that supports them, but from the
medium in which they grow.   When it is remembered
that the Diatomaceæ, and the Desmidiaceæ are both
included in Algæ, as well as the Confervæ of our
ditches, and the gigantic sea-weeds which form floating
banks in the ocean, it must be admitted that, in size
as well as appearance, they present an infinitude of
variety.

If we go back to the estimates of 1847 we shall find
that the total of Diatomaceæ was then estimated at
457 ; of the Confervaceæ, which included the Des-
midiaceæ, at 368 ; the Fucaceæ at 452 ; and the
Ceramiaceæ, almost equal to Floridæ, at 682, with the
Characeæ at 35 : making in all a total of scarcely
two thousand species for the whole world.   As affording evidence of the progress made in this department

of botany, we find that the latest work, in the form of a "Sylloge," gives the total of Diatomaceæ alone at 3272 species, with no less than 1118 Desmidiaceæ and 1860 for the residue of the Confervaceæ ; but even this is not complete, for one section of the Diatomaceæ is not yet published. Then we have to estimate the Floridæ, or red Algæ, so well known, by means of a few individuals, to the ordinary visitor to the seaside, as well as the dusky olive sea-weeds, or Fucaceæ, some of which attain extraordinary dimensions, and we cannot estimate less than a total of some ten thousand species.

With the exception of the fungi, the Algæ seem to be the most numerous in species of cryptogamic plants, although apparently not more than one-fourth of the number of fungi. The study of Diatomaceæ is often pursued by itself, and one great facility lies in the indestructible nature of their flinty valves, or skeletons. After being buried for centuries, enough still remains for their identification, and some of the most beautiful species are those dissolved out of the solid rock. It would be difficult to enumerate all the places where they are to be found, not only in salt water and fresh, but in all places where water has ever flowed, and even amongst the excrements of sea-birds. Possibly they are not absent entirely from any tract of country in the world, where there is or has been water. Of almost equal beauty are the Des-midiaceæ, but, as these are destitute of any flinty skeleton, they soon perish entirely, so that no trace of them can be found except in the living state ; more-over, for the most part they are confined to fresh or

G

brackish water.   Marine Algæ float in the seas
throughout the globe, and possibly a great number
may still flourish, attached to submerged rocks, which
have never met the eyes of man.   Altogether the
presumption might have been that Algæ would have
been more numerous in species than even the fungi,
since they inhabit a medium which covers a much
larger area, and, in the Diatomaceæ, possess facilities
for preservation, which no other cryptogams can
claim.

GIGANTIC SEA-WEEDS.

We cannot omit reference to the gigantic vegetation
of the ocean, which bears comparison with the " big
trees " of the land, and deserves a place amongst the
marvels of the vegetable world.   The foremost place
should perhaps be given to the Giant Fucus, which
was first noticed in the sixteenth century, and has
since been referred to by all voyagers in the south.
Captain Cook says that the stems attain a length of
one hundred and twenty feet.   That these dimensions
are considerably under the mark there is very little
doubt, though the report that specimens have been
measured upwards of one thousand feet is, perhaps, as
much an exaggeration.   Still, it has been observed
that, provided the water be smooth and of sufficient
extent, there are no impediments to the almost
indefinite elongation of the upper part of a plant
which never branches, and whose growth is independent of all below it, even of the root.   In Hooker's
" Botany of the Antarctic Voyage of the *Erebus* and
Terror," it bears the name of *Macrocystis pyrifera ;* and

it is stated of it that "specimens measuring between
one hundred and two hundred feet are common in the
open ocean, and these are always broken off at the
lower end, either from the division of the frond by
sea animals, through whose agency the plant increases
and the floating island it forms dilates, or from the
impossibility of securing the whole mass from the
motion of the vessel or the swell of the sea, in lati-
tudes where no boat can be lowered. Again, D'Urville, upon whose observations in natural history the
utmost reliance may be placed, states it to grow in
eight, ten, and even fifteen 'brasses' of water, from
which depth it ascends obliquely, and floats along the
surface nearly as far ; this gives a length of two
hundred feet. In the Falkland Islands, Cape Horn,
and Kerguelen's Land, where all the harbours are so
belted with its masses that a boat can hardly be
forced through, it generally rises from eight to twelve-
fathom water, and the fronds extend upwards of one
hundred feet upon the surface. We seldom, however,
had opportunities of measuring the largest specimens,
though washed up entire on the shore ; for on the
outer coasts of the Falkland Islands, where the beach
is lined for miles with entangled cables of *Macrocystis*, much thicker than the human body, and
twined of innumerable strands of stems coiled together
by the rolling action of the surf, no one succeeded in
unravelling from the mass any one piece upwards of
seventy or eighty feet long : as well might we attempt
to ascertain the length of hemp fibre by unlaying a
cable. In Kerguelen's Land the length of some of
the pieces which grew in the middle of Christmas

Harbour was estimated at more than three hundred feet ; but by far the largest seen during the Antarctic expedition were amongst the first of any extraordinary length which the ships encountered, and they were not particularly noticed, from the belief that the report of upwards of one thousand feet was true, or, at any rate, that better opportunities of testing its truth would arise in the course of a three-years voyage than the first week of our explorations could afford. These occurred in a strait between two of the Crozet Islands, where, very far from either shore, in what is believed to be forty-fathom water, somewhat isolated stems of *Macrocystis* rose at an angle of 45° from the bottom, and streamed along the surface for a distance certainly equal to several times the length of the *Erebus ;* data which, if correct (and we believe them so), give the total length of the stems as about seven hundred feet.

"That isolated patches of weed should rise through such a volume of water is not incompatible with the statements we have elsewhere made, that eight or ten fathoms is the utmost depth at which, judging by our experience, submerged sea-weed vegetates in the Southern Temperate, and Antarctic Ocean. These exceptional cases are probably due to the parent plant having attained such a size in its birthplace near shore, as to weigh its stony moorings, and deposit itself in deeper water, where an increase of the roots would unite the original base to other rocks, and thus gain a footing that defies the power of the elements." [1]

[1] Hooker, "Cryptogamia Antarctica," p. 157. (1845.)

This weed girds the globe in the southern temperate zone, but not in the tropics or northern hemisphere ; its southern boundary being very much
determined by the position of the ice, and its northern
boundary by the currents and temperature of the
water.

The above is not the only gigantic sea-weed which
is found in the Antarctic Ocean, since there is another
which vies with it in magnificence. This is *Lessonia*,
or, we should rather say, the species of *Lessonia*, for
there are two or three, which differ only in specific
details. " These are dichotomously branched trees,
with the branches pendulous, and again divided into
sprays, from which hang linear leaves from one to
three feet in length. The trunks usually are about
from five to ten feet long, as thick as the human
thigh, rather contracted at the very base, and again
diminishing upwards. The individual plants are
attached in groups, or solitary, but gregarious, like
the pine or oak, extending over a considerable
surface, so as to form a miniature forest, which is
entirely submerged during high water, or even half
tide, but whose topmost branches project above the
surface at the ebb. To sail in a boat over these
groves on a calm day affords the naturalist a delight-
ful recreation, for he may there witness, in the
Antarctic regions, and below the surface of the ocean,
as busy a scene as is presented by the coral reefs of
the tropics. The leaves of the *Lessonia* are crowded
with *Sertulariæ* and Molluscs, or encrusted with
Flustræ; on the trunks parasitic algæ abound,
together with chitons, limpets, and other shells ; at

the bases and amongst the tangled roots swarm thousands of the *Crustacea* and *Radiata*, whilst fish of several species dart amongst the leaves and branches. But it is on the sunken rocks of the outer coasts that this genus chiefly prevails, and from thence thousands of these trees are flung ashore by the waves, and, with the *Macrocystis* and *D'Urvillea*, form along the beach continued masses of vegetable rejectamenta, miles in extent, some yards broad, and three feet in depth ; the upper edge of this belt of putrefying matter is well in-shore, whilst the outer or seaward edge dips into the water, and receives the accumulating wrack from the submarine forests throughout its whole length. Amongst these masses the best algæ of the Falklands are found, though, if the weather be mild, the stench, which resembles putrid cabbage, is so strong as to be almost insufferable. The ignorant observer at once takes the trunks of *Lessonia*, thus washed up, for pieces of drift-wood, and, on one occasion, no persuasion could prevent the captain of a brig from employing his boat, and boat's crew, during two bitterly cold days, in collecting this incombustible weed for fuel." [1] The substance of the trunk is employed by the Gauchoes for knife-handles, and contracts in drying into a substance harder than horn.

The third remarkable alga of the Antarctic regions is *D'Urvillea utilis*. In Kerguelen's Land its enormous and weighty fronds, sometimes ten feet long, and almost too heavy for a man to lift, form the only shelter for the shells and soft animals, which

---

[1] Hooker, " Cryptogamia Antarctica," p. 152.

there find a refuge from the flocks of aquatic birds that cover the shores, and follow the receding tide.

One remarkable species of *Sargassum*, says Professor Harvey, "has long been famous by the name of Gulfweed, or Sargazo, under which most voyagers since the days of Columbus have spoken of it. That great discoverer was the first to encounter it in modern times, and with his account we are therefore most familiar; but possibly the weedy sea which Aristotle speaks of, as having been met with by the Phœnicians, at the termination of their voyage, may have been an early discovery of the same bank. It is curious that the great bank, which extends between the twentieth and forty-fifth parallels of north latitude, 40° west from Greenwich, appears to occupy the same position at the present day as it did in the days of Columbus. Between this bank and the American shores various smaller strata, and detached masses of sea-weed occur, being thrown into this portion of the ocean by the eddy caused by the sub-circular motion of the great oceanic currents. The whole of this immense space of ocean, which is reported to be thickly covered with sea-weed, is computed by Humboldt at upwards of 260,000 square miles, an area almost six times as large as Germany; but it is not to be supposed that all this space is *equally* clothed with floating verdure. In many places the weed occurs in distant and narrow ridges, leaving spaces of clear water between. This portion of the Atlantic seems to be the chief settlement of the *Sargassum* (Fig. 13), but straggling specimens occur in the Pacific and Indian Oceans, and on the shores of Australia and New Zealand ; and some few, carried northward by the Gulf Stream, reach the northern shores of Europe in safety.

"Naturalists have been puzzled to account for the origin of the Gulf-weed, and formerly it was supposed to be altogether de-rived from the Gulf of Mexico ; being torn off the shores of the Florida reefs and keys, and carried to sea with the great current. It is possible, and indeed pro-bable, that the origin of the present float-ing banks may have been partly of this nature, but it is most certain that the great masses of the weed that are at present found floating have had no such imme-diate parentage, but are produced on the surface of the ocean on which they float. Whoever has picked up the plant at sea, on any genuine portion of the bank, must have seen that it was in a perfectly fresh and growing state, and, if he

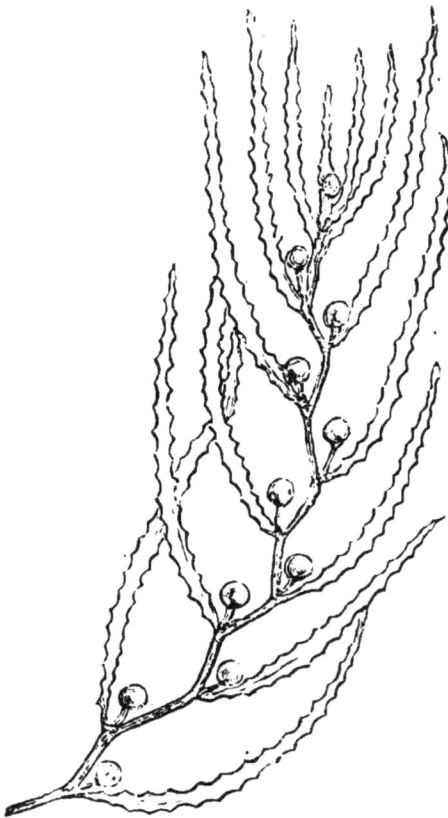

Fig 13.—Gulfweed, *Sargassum bacciferum.*

has looked at his specimen carefully, he will probably have observed that different parts of the same
specimen were of very different ages; that though
there was no apparent root, yet that toward the
centre of the mass a small portion of the stem was
of a much darker colour than the rest, and possibly
covered by parasitic incrustations, and that all the
branches, springing from this central piece, were
successively more and more delicate, and of paler
colour, and evidently in a young and sprouting state.
Such a specimen is clearly in vigorous life, yet it has
no *root*. But the absence of root is a matter of very
trivial moment in a sea-weed; for we must bear in
mind that the roots of Algæ are merely holdfasts,
intended to keep them from being washed off the
rocks on which they grow. And in a plant capable
of enduring extensive change of place, like this
Sargassum, the root is the part which may be most
readily dispensed with. No doubt the specimen
under examination originated in a little branch,
accidentally broken from a neighbouring mass, and
which, being thus cast adrift, continued to push out
new branches and leaves. In this manner, by the
continued growth of their broken parts, the floating
masses spread over the surface of the sea. In
this floating state the species never forms proper
fructification. There is therefore no growth from
spores. The supply of plants is consequently kept
up, and extended, by the constant development of
buds, or *gemmæ*, originating in broken fragments of
branches. This process of growth, by breakage,
must have gone on for ages, from that early time

when the first individuals were brought from some unknown rocks by the currents of the ocean."[1]

ACTIVE ZOÖSPORES.

No phenomenon in the cycle of plant life, as exhibited in its lower forms, is more instructive and suggestive than the movements of the sexual elements in some of the thread-like fresh-water Algæ, and especially those which bear the generic name of *Œdogonium*. In their simple and sterile condition, these Algæ consist of cylindrical threads, divided by transverse partitions into elongated cells. Whilst growing and vigorous these cells contain a green colouring matter, which imparts its colour to the threads, but in the course of time these contents are absorbed in the formation of the organs which are developed at their expense within the cells. In one series of species the little male plants are produced from other cells of the same thread as those which are converted into female cells. In another series the male plants are produced within threads distinct from those which develop female cells, so that some of the threads produce only female cells, and others only the male. Of the first series an example may be taken when in its mature condition, and then it will be observed that the threads are no longer of equal thickness throughout, but are swollen in some places to double their original diameter, in an abrupt and bladder-like manner, the inflations having a somewhat globose or elliptical form. The cells thus inflated are the vital female cells, which are termed

[1] W. H. Harvey, "Nereis Boreali-Americana," p. 53. (1851.)

oögonia, because they enclose the ovum, or egg-like body which is to become the fertilized oöspore, and reproduce the original plant. When the oöspore ultimately takes form it is globose, or elliptical, at first of a green colour, but finally brown, and lies freely within its mother-cell. This mother-cell, or oögonium, in addition to its inflation, becomes modified in another way, since it splits all round, above the middle, and the upper part divides from the lower, like the lid of a box, leaving a slit-like opening, through which communication from the exterior may be made with the spheri-cal body, or oöspore, nestling within (Fig. 14, *b*).

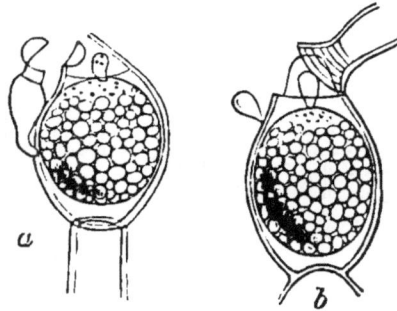

FIG. 14.—Oögonia ; *a*, with lateral pore ; *b*, with splitting operculum.

This is the most common modification ; but in rarer instances, instead of splitting all round and forming a lid, a perforation or opening is made in the wall of the mother-cell for external communication (Fig. 14, *a*). Hence it may be said that some oögonia open with a lid, and some are only perforated at the side by a pore. Whilst this process of development is going on in the specialized female cell, and it becomes changed into an inflated mother-cell, or oögonium, enclosing the incipient oöspore, or ovum, other cells are undergoing change upon the same thread, in order to become differentiated into active male cells. These latter are much shorter than ordinary cells, and may be single, or four to six

together, and are not inflated. Each of these cells
produces, in time, within its interior, a rounded or
elongated body, which at maturity breaks through the
cell wall, and escapes into the surrounding water (Fig.
15, *a*), endued with all the attributes of active life.   In
this condition it is called an *androspore*, literally a male
spore, furnished with delicate cilia, or movable hairs,
at one end, by means of which it moves freely in the
water, seeking some place at which to come to rest (*b*).

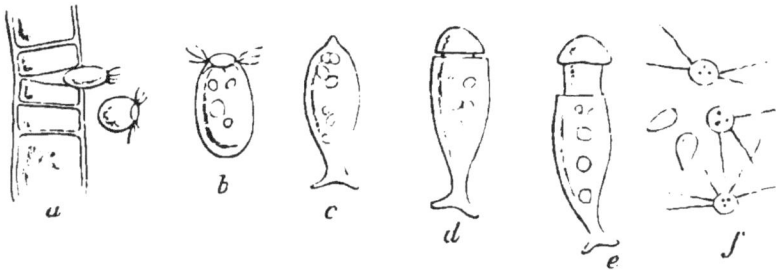

Fig. 15.—Growth of androspore.

When the desired resting-place is found the free
androspore attaches itself by the fringed end, and
thenceforward becomes a fixed object.   When this
is accomplished a further growth takes place, as it
acquires a flask-like shape, attenuated downwards
into a more or less elongated stem (*c*), and the extreme
apex narrowed into a mouth, or opening, closed by
a deciduous lid.   The androspore becomes converted
into a little male plant, or dwarf male, attached by
its base to some portion of the original thread (*d*), and
containing within itself certain fertilizing elements,
the spermatozoids, which escape, when mature, at
the apex, by the falling away of the lid, in order to
fertilize the oöspore, contained in the above-named

inflated cells, by entering the pore, or by the slit caused by the raising of the operculum (e). It is at this juncture that the most curious phenomenon takes place, which has all the appearance of intelligent action. After the androspores escape from their shortened mother-cells, they float freely in the water, in search of some position in which to attach themselves, for further development (f). Their aim and object would be, if they were free and intelligent beings, to attach themselves as near as possible to the oögonia, which they are destined to fertilize, in order that their contents might be discharged into the openings left for that purpose, otherwise the spermatozoids would be diffused in the surrounding fluid, and fail of their mission. The true facts of the case are, that the androspores *do* attach themselves upon, or immediately around the inflated cells, and there attain their final development, in close proximity to the openings which the mature spermatozoids are destined to enter. It would be folly to ascribe this to mere accident, otherwise androspores would commonly be detected seated on barren portions of the threads, far away from the inflated cells ; but, on the contrary, they will always be found clustered about the oögonium, in just such a position as that in which they would be expected to place themselves, had they been guided by the same instinct which controls the actions of birds and animals. Nothing more suggestive than this occurs amongst all the phenomena of low life, and no explanation has yet solved the mystery, which is intensified when we inquire into the history of the life of the other species, in the same genus, in which the androspores are *not* produced from the *same* threads as the female cells.

In this second group, the vegetative threads are uniformly of two kinds ; there are threads which produce androspores, from shortened male cells, and other threads which produce only the inflated female cells, or oögonia. The two kinds are never present on the same thread. The processes of development are the same, and the phenomena are the same, with the exception that the androspores, when they escape from the mother-cells into the surrounding fluid, have to seek out the female threads first, and then afterwards discover the portion of those threads which bears the inflated cells.[1] There is every appearance, in such a process, of a power of selection which we can scarcely believe to exist in simple plants, not accredited even with instinct. But the facts are indisputable, the selection is made, the escaped androspores find their way to the female threads of their own species, and they attach themselves to those threads, in the immediate vicinity of the inflated cells

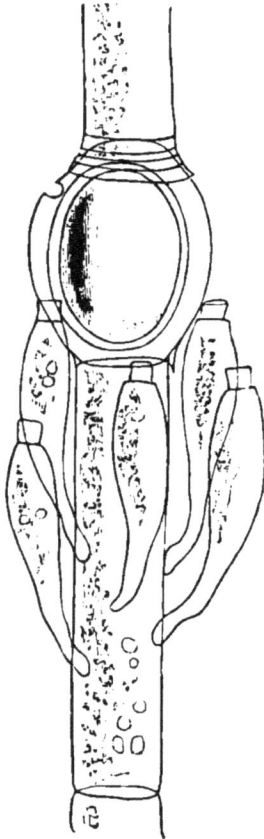

FIG. 16.—Oögonium, surrounded by dwarf males attached.

---

[1] M. C. Cooke, "British Fresh Water Algæ," p. 149 (1882); "Introduction to Fresh Water Algæ," p. 145 (1890).

(Fig. 16), but they do *not* attach themselves indiscriminately to other objects, or to other threads belonging
to other species, or even to the sterile, or the male
threads of their own species. Let theorists explain
the phenomena as they will, it must be an intelligible
theory, capable of accounting for the results, for
chance or accident must be entirely outside the
calculation. The development of active zoöspores
is not at all an uncommon event in the life history
of Algæ, and, when first discovered, led to a great
deal of discussion, but in no instance are they
accompanied by such startling phenomena as are to
be met with in *Œdogonium*, and in a less degree in
the allied genus *Bulbochæte*.

It was at one time strongly suggested, if not
insisted upon, that the articulated Confervæ, and
especially the Conjugatæ and the *Œdogoniaceæ*, had
intimate relations with the animal kingdom. The
discovery of the mode of reproduction in such genera
as *Œdogonium* led Mr. Hassall to remark that "it
throws much light upon the often-canvassed and
much-disputed subject of the animality of the conjugating genera. It proves, since in reality a conjugation is necessary to the formation of every true spore,
that all the Confervæ stand upon the same footing as
regards their animal nature, and that if those species
which exhibit the curious phenomena of conjugation
are really animal, so are all the other Confervæ
mentioned ; that if these should ever, at any
subsequent period, be removed from the vegetable
kingdom to the animal, so ought, as a *sequitur*, all the
other Confervæ alluded to, the *Œdogoniaceæ* and the

Sphæropleæ. But it appears to me that the facts thus disclosed, so far from adding weight to the arguments of those who would regard the Conjugatæ as animal productions, rightly interpreted, tend merely to prove the existence of sexes in the Confervæ, as have been proved by Vaucher to exist in the genus *Vaucheria,* a class of productions nearly related to the Confervæ; and that thus an analogy is established between the lower Confervæ and the higher phanerogamic plants, between which and some of the lower animal tribes a further analogy may be traced.

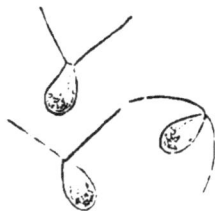

Fig. 17. —Zoö-pores.

For my own part, I trouble myself but little with the disputes about the boundaries of the two great divisions of the organized world, which forcibly remind me of the search, carried on by ancient philosophers for days and years, after the much-desired but imaginary and poetical philosopher's stone endowed with such all-pervading influence, or the equally fruitless inquiry after perpetual motion, or any of the other wild chimeras to which the minds of men have from time to time been given. It is my belief that no such rigid boundary exists, for in living nature there are no abrupt unsightly chasms; all is uniformity, order, design, and transition. I would now mention one fact which would appear to show that in the composition of the Confervæ there is something of the animal. When a number of Confervæ have been crowded together in a bottle for two or three days, they emit on their removal what appears to my power of smell to be a strong animal and offensive odour. A similar offensive smell is emitted by some marine sponges in a state of decay."[1]

It hardly need be urged that the final reason is an inconclusive one, and that all the foregoing are based upon the assumption that conjugation, and spontaneous movements, are exclusively zoological phenomena, which is now admitted to have been a cardinal error.

MOOR BALLS.

The first notice discoverable of the occurrence of "moor-balls" in this country is that contained in the *Philosophical Transactions* for the year 1751, when they were found in Yorkshire by Mr. William Dixon. "They were taken up in a fresh-water lake on a large common in the East Riding of Yorkshire, about twelve miles west of Hull. The lake is from one hundred to two hundred acres in size, according to different seasons, and empties into the Humber, which is pretty salt, and has sometimes infected it a little at very high tides. The water is very bright, and the bottom in many places is quite covered with these balls, like a pavement, at different depths. These now sent were about six inches under water, and many are left quite dry every summer." On this communication Mr. Watson observes that the vegetable here mentioned he had never seen before ; neither had he been able to find it described in any of the botanical writers he had consulted. The matter of which it is composed is that of a Conferva, and should therefore

[1] A. H. Hassall, "Notices of Fresh Water Confervæ," in *Annals and Mag. Nat. Hist.* (Dec., 1842), vol. x. p. 336.

have had a place under that genus in Dillenius. The balls are of a deep green mossy colour, are hollow, of an irregularly spherical figure, and of different sizes, from an inch and a half to three inches in diameter. They are covered with very short villi externally, and the thickness from their external to their internal surface is about a quarter of an inch ; their texture is most compact the nearest to the surface. He denominates them globose conferva.[1]

Subsequently it was discovered that these globose masses of algæ were not wholly unknown, since the plant had been described by Linnæus as *Conferva ægagropila*, under which name it came afterwards to be included in British Floras, and was recorded from several localities. The great peculiarity is its growth in compact balls, often as large as a cricket ball. The "very short villi" with which the balls were said to be covered externally were simply the short free ends of the alga, of which the balls are wholly composed. When the plant came to be figured in " English Botany " (pl. 1377) in 1804, it had been found in North Wales and in Shropshire, and all known information, up to date, was collected. "They are the growth," it is stated, " of Alpine lakes in many different countries, and lie in great abundance at the bottom of the water. Their size is from that of a pea to three or four inches in diameter, and their form always pretty exactly spherical. Internally they are hollow, and quite destitute of any nucleus. When separated they are found to consist of innumerable green pellucid jointed

[1] W. Dixon, " On Some Vegetable Balls," in *Philosophical Transactions of the Royal Society* (1751), vol. xlvii. p. 280.

filaments, repeatedly branched, and firmly entangled together. The joints contain a green fluid substance, which by drying settles in an opaque form at their extremities, as in others of the genus. No traces of real fructification have been observed, though the extreme points of the filaments have an appearance which might be mistaken for such. It would seem that several of these filaments spring from one centre, perhaps fixed to some earthy particle, which, like the Dodder, they soon leave, and their lower parts wither away, while by branching and extending themselves upwards they form a gradually enlarging globe." Mr. Williams informs us of these balls being used to wipe pens upon. The specific name alludes to their resembling the hairy balls found in the stomachs of goats.

Hassall mixed up the present species with another, *Cladophora glomerata*, because he considered them merely as conditions the one of the other. The compact form, which constituted the " moor balls," he believed to be formed as follows : " A specimen, by the force of some mountain stream, swollen by recent rains, becomes forced from its attachment ; as it is carried along by the current, it is made to revolve repeatedly upon itself, until at last a compact ball is formed of it, which finally becomes deposited in some basin or reservoir in which the stream loses itself, and in which these balls are usually found."[1]

The appearance which these balls present is not by any means suggestive of an agglomeration of filaments

[1] A. H. Hassall, "A History of British Fresh Water Algæ," vol. i. p. 215. (1845.)

which have been entangled together and rolled into a ball, but of a plant growing in a dense mass, with the filaments all of one kind, and disposed centrifugally in one direction. At first the dense tuft may have been attached to some matrix, at length becoming free, and then by equal growth outwards obliterated all trace of attachment, and assumed the spherical form. Smaller and less solid tufts are conglobated by other free-swimming algæ, belonging to other genera.

What is Nostoc?

Let us introduce this question by a quotation, which gives, within a moderate compass, an explanation of the problem. "Suppose, for instance, the student, after a few hours' train, goes out into the open air, and sees the gravel and short grass strewed with gelatinous puckered olive-coloured masses, of which he perceived no trace a few hours before, his curiosity is excited, and he is anxious to ascertain the nature of this pro-duction. Externally it presents no marked differences, and within it seems to consist of a uniform jelly, without anything to make him suppose that it can be a mass of eggs. He examines it under the micro-scope, and finds that it consists of necklace-like chains of pellucid granules immersed in jelly of no definite structure. Some of these are larger than the others. He finds after a time that they change colour and increase considerably in size, though still retaining a regular outline; presently, the matter contained in their cavity becomes organized, and a new necklace of spores is contained within it : in fact, he has a young

repetition of the perfect plant, requiring only extension
of parts to assume completely its size and aspect. This
answers to the first part of the definition of a crypto-
gam ; but the plant does not germinate, he can discover no sexual indications, though germination does
not take place by the protrusion of a filament, and the
protoplasm of the cell at once gives rise to a new
plant. He believes it to belong to the vegetable
kingdom, and he feels that he has hit upon one of
those exceptional cases which defy mathematical accuracy. But still he has no doubt about the matter.
The plant is *Nostoc commune* (Fig. 18), a widely dis-

FIG. 18.—*Nostoc commune.*

tributed alga, bordering very close upon the gelatinous
lichens." [1]

There are several species, but this is the commonest,
and on some accounts the most interesting, especially
in its habitat and in the manner of its appearance.
Curious notions have been current as to its origin,
since some have supposed it to fall from the clouds in
a shower of rain, which may have originated in its

[1] Berkeley, "Introduction to Cryptogamic Botany," p. 16.

sudden appearance on garden paths, in masses occasionally as large as a walnut, after a shower. Some of the species are found in rivulets and ditches, some on the ground, or on rocks, and some on walls or glass. All are like little lumps of jelly, and when examined microscopically are found to contain chains of globose cells, variedly curved, immersed in the gelatine (Fig. 19). By a singular coincidence this same kind of structure is found in the thallus of the gelatinous lichens, of the group *Collemaceæ*. There have been theorists who have contended that the alga is only an

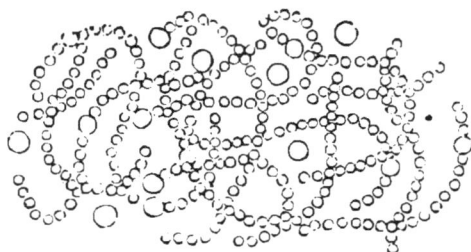

FIG. 19.—Cells of Nostoc.

early and imperfect condition of the lichen, and we are reminded of certain experiments, in which it is assumed that, by inoculation of the gelatinous mass of the *Nostoc*, the *Collema* has been produced, artificially, with the true fructification of a lichen. Probably something of this kind first suggested, or gave countenance to, the theory, elsewhere alluded to, of the dual nature of lichens.

Normally the trichomes, or chains of subglobose cells, in Nostoc have either terminal or intermediary cells, of a different colour and slightly different size, which are termed *heterocysts*, but their function is somewhat obscure. New plants are originated from fragments of the trichomes, which are called *hormogones*. The mucilage of old plants becomes softened, and then the fragments of the trichomes escape, leaving the heterocysts behind. The escaped fragments, or hormogones, become endowed with motion, similar to that observed in the threads of *Oscillaria*. The cells of the hormogones increase by division at right angles to the filament, ultimately separating longitudinally, and becoming the centre of new plants. Thus it will be seen that the Nostoc has in itself the power of reproducing the species by means of the hormogones, as well as by another process, which is the conversion of certain privileged cells into spores, these latter germinating, and giving origin to new plants.

Brief allusion may be made to the theories linked with the Nostocs by different authors. Professor Sachs has recorded that he watched a quantity of *Nostoc commune*, and saw it develop into the lichen called *Collema pulposum*. He states that the peculiar collema threads first appeared as little lateral offshoots from the cells of the nostoc filament, and rapidly developed into well-formed collemoid filaments. He says that every possible stage from the typical Nostoc to the typical Collema was seen repeatedly. These observations have, however, never to our knowledge been confirmed.

Another author, M. Baranetsky, professes to have seen the reverse process of Nostoc developed out of a Collema. He placed sections of actively growing fronds of *Collema pulposum* on smooth damp earth, using proper precautions to prevent external influence. After some days the sections became less transparent and intensely green from the crowding of the gonidia, which were now arranged in curved rows, closely rolled together into balls. Upon the upper surface of the section appeared little gelatinous balls, or warts, which contained gonidia in rows, and gradually developed typical Nostoc forms ; whilst on the edges of the sections appeared little colourless wart-like masses of jelly, in which, after some time, appeared gonidia, some of which developed into the typical Nostoc forms, others into true collemoid plants. He further states that he watched the body of the section gradually change, by the continual growth and increase of the rows of gonidia alluded to, and by the disappearance of the collemoid threads, until at last the whole mass of the tissues of the lichen had been converted into a true Nostoc, which was finally identified as *Nostoc vesicarium* (D.C.). Thus it will be observed, if both the above accounts are to be relied upon, *Nostoc commune* produces *Collema pulposum ;* and *Collema pulposum* originates *Nostoc vesicarium*, and not *Nostoc commune*.[1]

BOILED ALGÆ.

Amongst the most beautiful of the fresh-water algæ which are found in our ditches and streams, are the different species of *Batrachospermum*, which are great favourites with all young microscopists (Fig. 20). The number of species is dependent very much upon individual opinion as to the limits of species, some botanists believing in a large number of species, others in but a small number of species, with a great

---

[1] Further observations on this subject, by Dr. Braxton Hicks, will be found in the *Quarterly Journal of Microscopical Science*, 1861, p. 90.

tendency, and almost unlimited power of variability. Whatever the number recognized, they are all gelatinous, so that, when removed from the water, they resemble as much the spawn of some batrachian, as they do a water-plant. Under the microscope they are characterized by whorls of beaded threads, which vary according to the species, but barren description would fail to describe their beauties, which can only be realized from faithful figures.[1] The most remarkable feature about these algæ is perhaps the incident narrated by Bory. "I made," he writes, "twelve years ago, an experiment which ought to be known.

FIG. 20.—*Batrachospermum*.

After having many times carried from one locality to another stones bearing individuals of this species, which continued to prosper in spite of the change of habitation, I steeped many of them in lukewarm water, afterwards in boiling, and no part of the Batrachosperm appeared under the microscope to have undergone the slightest disorganization by these immersions, and certain sprigs replaced in their native place continued to vegetate after these experiments. I do not think that there exist other vegetables which boiling water does not immediately

---

[1] See the various species and subspecies figured in "British Fresh Water Algæ," by M. C. Cooke, pl. 120 to 126. (1884.)

disorganize; there are not others that can resist temperatures so opposite." Admitting the truth of this observation, which we are not certain has been confirmed, it is probably due to the gelatinous investment which protected the cell contents from injury, since it is known that a mucous or gelatinous investiture of vegetable cells is a great protection against the extremes of heat and cold. Undoubtedly some species of *Oscillaria* will flourish in thermal springs, and in currents of water so hot that the hand cannot be held in them, as evidenced in the waste from steam-engines in manufacturing districts, which practised algologists will always examine in search of rare species of *Oscillaria*.

A species of *Nostoc*, also a gelatinous alga, has been described by H. C. Wood, which was found in California in a thermal spring. "In the basin are produced the first forms, partly at a temperature of 124° to 135° Fahr. Gradually in the creek, and to a distance of a hundred yards from the springs, are developed, at a temperature of 110° to 120° Fahr., the algæ, some growing to a length of over two feet, and looking like bunches of waving hair of the most beautiful green." This alga was named provisionally *Nostoc calidarium*;[1] mixed with this was a species of *Chroococcus*, an unicellular alga, consisting of a few cells, surrounded by a mass of transparent firm jelly. In both cases the algæ were of a distinctly gelatinous character, the investment preserving them from the injurious effects of the hot water.

In 1872 Dr. Blake reported that he had collected

[1] *Annals of Natural History*, September, 1868, p. 231.

diatoms in a hot spring in Pueblo Valley, Nevada, the temperature of which was 163° Fahr. More than fifty different species were recognized by him, and they were found to be mostly identical with the species found in the beds of infusorial earth in Utah, described by Ehrenberg, showing that the latter must have been accumulated in a hot lake, of about the same temperature. No other living species were found in the hot waters, excepting red algæ.

In one of the hot springs at the California geysers, having a temperature of 198° Fahr., he found two kinds of confervæ. In another spring, with the temperature at 174° Fahr., many *Oscillariæ* were found, which, by the interlacing of their fibres, formed a semi-gelatinous mass ; and in the water of a creek, at a lower temperature, the algæ formed layers sometimes three inches thick, covering the bottom of the pools, mixed with the same diatoms as were found in the 174° spring.[1]

The *Challenger* expedition obtained algæ from the boiling and hot springs in the Azores, amongst which *Botryococcus Braunii* was one of the most common forms, mingled with some *Chroococcus* and species of *Oscillaria*. Dr. Hooker found two species of *Confervæ* at the hot springs of Belcuppee, in the Behar hills, in broad luxuriant strata, wherever the temperature was cooled down to 168° Fahr.

ACTINOPHRYS.

Some remarkable phenomena were described by

[1] *Annals of Natural History*, October, 1872, p. 312.

Mr. H. J. Carter in 1856,[1] as having occurred in the cells of *Spirogyra crassa*, a large species of fresh-water alga. "Under certain circumstances," he says, "the cell of *Spirogyra* apparently dies, the chlorophyl becomes yellow, and the protoplasm, leaving its natural position, divides up into portions of different sizes, each of which encloses more or less of the chlorophyl; these portions travel about the cell under a rhizopodous form, the chlorophyl within them turns brown, the portions of protoplasm then become actinophorous, then more radiated, and finally assume the figure of *Actinophrys*. The radii are now withdrawn, while the pellicle in which they were encased is retracted and hardened into setæ with the rest of the pellicle, which now becomes a lifeless transparent cyst; another more delicate cyst is secreted within this, and the remains of the protoplasm within all having separated itself from the chlorophyl, divides up into a group of monociliated monads, which sooner or later find their way through cysts into the cell of the *Spirogyra;* while the latter by this time having passed far into dissolution (not putrefactive) they thus easily escape into the water. Putrefactive decomposition at the commencement destroys this process altogether.

"At first it did not appear plain why the portions of protoplasm enclosed the chlorophyl, but afterwards it was found that this was for the purpose of abstracting the starch which accompanies the latter, since in some cases where the grains of starch were

---

[1] "On the Change of Vegetable Protoplasm into Actinophrys," by H. J. Carter, in *Annals of Natural History* (1857), vol. xix. p. 259.

numerous the chlorophyl was not included. This
was the process when the cells of *Spirogyra* were
not pregnant with starch, as they are just before
conjugating. When these changes took place at this
period they were somewhat different, insomuch as
the whole of the contents of the two conjugating cells
became united into one mass, and having assumed a
globular form, remained in that state until the chlorophyl had become more or less brown. After this the
protoplasm reappears at the circumference of the
mass in two forms, viz. in portions which leave
the mass altogether, after the manner of rhizopods,
and in the form of tubular extensions, which maintain
their connection with the mass throughout. In both
instances the protoplasm is without chlorophyl, but
charged with oil-globules, and both forms make their
way to the confines of the *Spirogyra*-cell, which they
ultimately pierce, develop their contents, and discharge them, in the following manner :—

"On reaching the cell wall, each form puts forth
a minute papillary eminence, which, having passed
through the wall, expands into a large sac, or bursts
at its apex. Following the isolated form first, this
then gradually drags four-fifths or more of its bulk
through this opening, sometimes so much as only to
leave a little papillary eminence in it, which then
makes the portion of protoplasm look as if it were
entering instead of escaping from the *Spirogyra*-cell ;
the internal contents of this protoplasm then become
more defined and granular, the granules assume a
spherical form respectively, they evince a power of
locomotion, and the originally flexible pellicle having

become a stiffened cyst, with a more delicate one within, assumes a slightly conical form, which, giving way by a circular aperture at the apex, allows the granules to pass into the water, when they are seen to be monociliated monads ; each consisting, apparently, of a film of protoplasm expanded over an oil-globule, and bearing a single cilium. The contents of the tubular form, on the other hand, undergo the same changes, but the tube becomes dilated in a pear-shape within the *Spirogyra*-cell ; and when the monads are ready to lead an independent existence, the end of the papillary eminence, which has been projected some little distance beyond the cell wall into the water, gives way, and thus they also escape.

" In another form of this tubular extension, the inner delicate cyst expands into a flask-like or globular shape beyond the papillary eminence, outside the cell wall, and retains the protoplasmic contents here until they are ultimately developed into monads. These, which are much larger than the monads developed by the other processes, on issuing move about rapidly for some time, by the aid of a strong cilium carried in front, like that of *Astasia*, and then become stationary ; the 'contracting vesicle' which does not appear before they leave the cyst, now becomes very active, the cilium is gradually diminished in size and altogether disappears, and the monad passes into a rhizopodous, reptant state, which afterwards becomes actinophorous, and finally assumes a form undistinguishable from that of *Actinophrys sol.*"

Up to this point Carter had been able to follow this

transformation, and, although he had not actually
seen the actinophorous form enclose particles of food,
yet he deemed the form itself sufficiently significant
to guarantee this induction, since he had never witnessed a rhizopod of this kind without its attacking
everything, living or dead, that it could overcome
and turn into nourishment; besides, such a form
could obtain sustenance in no other way. If this
was not satisfactory, it was not difficult to conceive
that what the portions of protoplasm in an actinophorous form would do within, they would do
outside the cell of *Spirogyra;* and it had been
shown, in the first process detailed, that inside the
cell they enclosed chlorophyll, and finally ejected
the refuse in the manner of *Amœba.*

It was true that the transformation of the protoplasm of the cell of *Spirogyra*, and its movements
above detailed, were unlike the phenomena of vegetable life, but the formation of the spore itself in the
normal way, and the movements of the protoplasm
of the conjugating cells just preceding it, merely
required to be studied to bring about the conviction
that one was but a modification of the other. In the
abnormal way the chlorophyl died, two cysts were
formed around the portions of protoplasm respectively, the starch passed into oil, the refuse of the
chlorophyl was thrown off from the enclosed protoplasm in the manner of a rhizopod, the protoplasm
divided up into monads, which came forth as animals
—that is, in the form of rhizopods endowed with the
power of locomotion and polymorphism, and thus
under a form which does not live by endosmosis,

but by the enclosure of crude material from which the nutriment is abstracted by a digestive process, and the refuse finally discharged.

Lastly, whenever a mass of filaments of *Spirogyra* underwent these transformations, the latter were invariably followed by a numerous development of *Actinophrys sol* of all sizes, to the exclusion at first of almost all other animalcules; and, coupling this with the undistinguishable form from *Actinophrys sol,* assumed by the monads developed by these transformations, he saw no other more reasonable conclusion to come to, than that they were one and the same, and therefore that one source at least of *Actinophrys sol* was the protoplasm of *Spirogyra.*

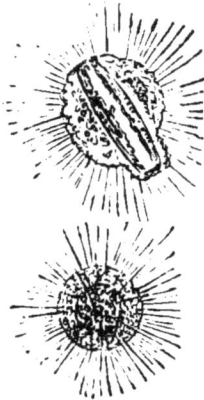

Fig. 21.—*Actinophrys sol.*

Water Blossom.

Under this name a peculiar phenomena is known in Germany, which in this country has been called the "breaking of the meres." Professor Cohn observes that, though the appearance has often been observed and examined, very little is known of the causes from which it originates. Within the course of a few hours an alga so densely covers a vast extent of the surface of the water that it imparts to it a distinct colour, green, brown, or red; sooner or later it disappears, either periodically or altogether. The only reason for this that can be assigned, apart from the extraordinary increase of the respective species, is the sudden change of their specific gravity, which causes them to rise suddenly from the bottom of the water, where they are developed in vast numbers, to the surface, and as suddenly sink down again. He instances the river Leba, near the Prussian frontier, and observes that " this Leba is a true moor river ; its banks are quite flat, the bed is nothing but moor and swamp, which gives way under one's feet. Whenever the river is about two feet deep, the water takes a brown colour, which prevents people from seeing to the bottom.   On July, 1877, the river appeared quite green, from a vast quantity of minute spherical bodies which floated on its surface, and even ordinary people were struck by it.   The phenomenon, which was first noticed towards noon, lasted for about five hours, and had totally disappeared in the evening.   The next morning there was nothing to be seen, but at noon there was again a large quantity, whilst there were very few towards night.   It was similar on the third day ; but since then the minute spherical bodies have entirely disappeared from the Leba." [1]   There does not appear to have been any further determination of these spherical bodies, than that they received the temporary name of *Rivularia fluitans*,[2] which, it is just possible, was the same as the *Rivularia* hereafter named.   The phenomenon was much more transient than in the case of the Shropshire meres.

A report made upon what was called the "breaking

[1] M. C. Cooke, Introduction to " British Fresh Water Algæ," p. 167. (1890.)

[2] *Hedwigia* (1878), vol. xvii. p. 1.

I

of the waters" narrates that this phenomenon occurs at certain seasons of the year in Ellesmere and some of the other Shropshire meres, which the people in the neighbourhood are accustomed to call "breaking of the water," or "breaking of the mere." To a stranger these terms are somewhat misleading, as they appear to suggest a violent agitation of the water, or its bursting through its banks, whereas the phenomenon resembles the breaking of wort in the process of brewing, causing a discoloration of the water, rendering it unfit for consumption, and spoiling the fisherman's sport. In its normal condition the water is pure and limpid, perfectly suitable for domestic purposes, but when it "breaks" it becomes turbid, from the formation of small dark-green bodies, in countless thousands, which not only float as a scum on the surface, but abound throughout the whole of the water. The change is so apparent that it cannot escape the notice of the most careless observer. On examining the floating matter of Ellesmere, the green bodies composing it are found to be rather smaller than a turnip seed, spherical in form, and of the deep green colour familiar to us in the rust of copper. Their specific gravity must be nearly the same as that of the water, which will account for their rapid dissemination throughout it when disturbed, and rising to the surface when at rest. This mere abounds in fish, and is much frequented in the proper season by anglers, but as soon as the breaking begins all sport invariably ceases, and the fish become torpid, refuse the bait, and sulk at the bottom. Whether this curious effect upon them is caused by some injurious gases generated at the time, or by the minute green bodies already mentioned entering their gills and impeding respiration, is a question not yet determined.

Various popular explanations have been given of this breaking, the more generally accepted one being that it results from the seeds of aquatic plants, growing on the margin of the mere, falling into the water ; and there is some probability on the face of this explanation, because it generally occurs in the autumn, when plants begin to drop their seeds, and the green bodies somewhat resemble a minute seed. In 1878 it was intimated that the real cause was the rapid germination of a minute plant classed amongst the algæ, and formerly known as *Conferva echinata*, but which it was suggested should be called *Rivularia articulata* or *Rivularia echinata* (Fig. 22).

It is necessary to observe that the phenomenon called " breaking " must be distinguished from a turbid or muddy state of the water, produced by heavy rains washing down vegetable fragments and earth. If we examine water under the microscope, changed in its appearance by this latter cause, we do not find one or two small vegetable organisms pervading the whole body of the water, imparting to it their own peculiar colour, as in true "breaking." Nor must we confound with it an occasional and partial occurrence of algæ in small quantities, for at any time during the year interesting species of these minute plants can be found, by diligent searching, in nearly every gently running stream, quiet pool, and mere. It can be most readily detected by the uniform dark-green colour of the water, or by the floating

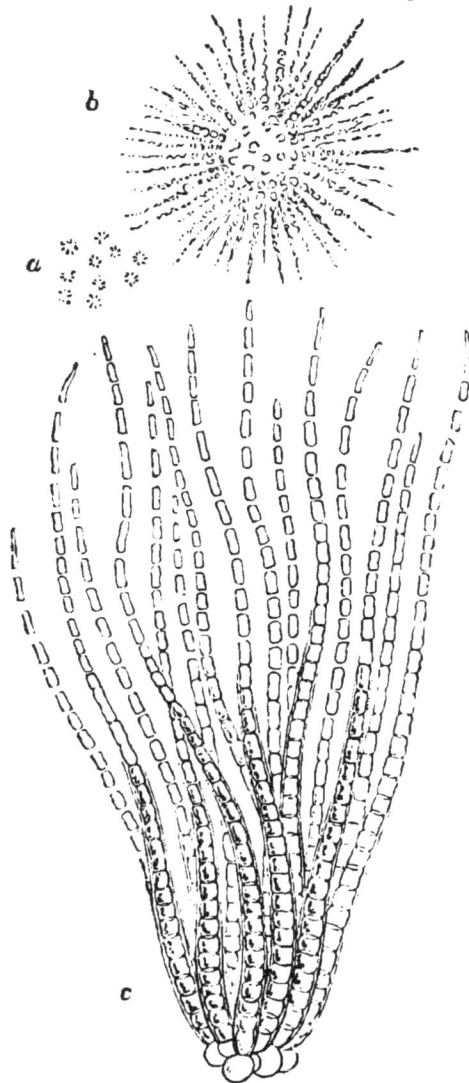

FIG. 22.—*Rivularia echinata.* *a*, natural size; *b*, enlarged; *c*, portion
magnified 400 times.

scum in the quiet bays on the leeward shore; but in

such cases it is best to take up a small portion in a white glass bottle, and look through it with a pocket-lens, when well-defined forms will be detected, though too small to admit of their structure being seen. A good microscope will at once show the myriads of beautiful green bodies — true plants — which are present. To convey some idea of their number, Mr. W. Phillips took a common pin, put the head of it in water collected in Newton Mere, and thus obtained a small drop, and, on placing this drop under a micro-scope, he could clearly count three hundred individual plants. It must be left to the reader to calculate how many must be present to colour the water of a mere one hundred and fifteen acres in extent. It only remains to add that the particular species of algæ found in the Shropshire meres at the time of "breaking" were the *Rivularia echinata*, in great quantity, one or two forms of *Anabæna* with *Aphani-zomenon flos-aquæ*, just as in Scotland, but with the addition of another minute species, *Cœlosphærium Kutzingianum*. We have yet to ascertain the reason for the sudden and enormous influx of the *Rivularia* and *Anabæna* at this particular season of the year, and its persistency for a long period in Britain, whilst so transient in Prussia. Here again, as in so many other instances amongst the lower forms of vegetable life, we recognize a deficiency in our knowledge of the life-history of these minute organisms; whereas, if we were able to trace them through their whole career of development and reproduction, we might comprehend, and appreciate at its true value, as a perfectly natural consequence, this sudden appearance of myriads of a minute green alga, of which not less than three hundred could be contained in a single drop of water on the head of a pin.[1]

Professor Dickie has placed on record his experiences of a similar phenomenon in Scotland, in which almost the same organisms were concerned. He writes, "For some years back excursions have been made with the students of my botanical class to a loch on the estate of Parkhill, about four miles north-west from Aberdeen. The sheet of water in question is about a quarter of a mile in its greatest length; on almost all sides surrounded by extensive deposits of peat, with the soluble matter of which a great proportion of the water passing into the loch is impregnated. The locality was generally visited in the beginning of July. Nothing peculiar had ever been observed till the summer of 1846, when my attention was arrested by a peculiar appearance of the water, especially near the edge, but extending also some distance into the loch. Numerous minute bodies, with a spherical outline, and varying in size from one twenty-fourth to one-twelfth of an inch in diameter, were seen floating at different depths, and giving the water a peculiar appearance. In some places they were very densely congregated, especially in small creeks at the edge of the loch. A quantity was collected by filtration through a piece of cloth, and, on examination by the microscope, there could be no doubt that the production was of a vegetable

---

[1] "The Breaking of the Meres," by W. Phillips, in *Transactions of Shropshire Natural History Society*. *Grevillea*, 1880, p. 4; 1882, p. 111.

nature, and a species of *Rivularia*. It was afterwards pronounced to be *Rivularia echinata*. Along with it, but in very small quantity, was *Anabæna flos-aquæ* (Fig. 23).

"In the first week of July, 1847, the same species was observed similarly associated, but the *Anabæna* was now more plentiful, without any corresponding diminution in the quantity of the *Rivularia*. In July, 1848, it was observed that the *Rivularia* was as rare as the *Anabæna* had been in 1846 ; to the latter, consequently, the water of the loch now owed its colour, which was of a very dull green : the colour, however, becomes brighter when the plant is dried. In neither of the seasons mentioned was it in my power to make any observations on the colour of the loch earlier or later than the date above mentioned, consequently nothing can be added respecting the comparative development and progress of the two plants at other seasons."[1]

Fig. 23.—*Anabæna flos-aquæ*.

De Candolle has called attention to Lake Morat, in Switzerland, stating that during every spring it presents the appearance of a reddish scum upon its surface, which the fishermen express by saying that the lake is "in flower." In 1825 the phenomenon continued from the month of November until May, and its unusual exuberance was supposed to originate

[1] *Annals and Magazine of Natural History*, January, 1849, p. 21.

from the great mildness of the winter, the consequently smaller elevation of the water of the lake being also favourable to the development of the matter, evidently organic, which caused the redness.

During the early hours of the day the lake presented nothing remarkable, but soon after there appeared long, red, very regular and parallel lines along its borders, and at some distance from the shore; the breezes urged this matter into the little creeks, and heaped it up around the reeds. There it covered the surface of the lake like a fine reddish scum, forming patches of colours varying from greenish black to a beautiful red. It was also seen of every shade of yellow, red, and grey; some of the patches were marbled, and others presented figures much resembling those produced by positive electricity on the electrophorus. During the day this mass exhaled an infectious odour, but during the night all disappeared, to be renewed on the following day.

When the lake was agitated by strong winds the phenomenon disappeared, but again presented itself on the re-establishment of a calm. Many species of fish, as perch and pike, probably from having eaten of this matter, had their bones and flesh tinged red, as if they had been fed on madder, but without any inconvenience. The particular alga in this instance was called *Oscillaria rubescens*.[1]

A similar experience is related by P. H. Gosse as having occurred to him within a very limited space,

---

[1] De Candolle, in *Mémoires de la Société de Physique et d'Histoire Naturelle et Génève*, vol. iii. p. 29; *Ann. Nat. Hist.*, vol. i. p. 4.

that is to say, in a tank of sea-water. "Patches of a rich crimson purple colour formed here and there on the surface, which rapidly grew on all sides till they coalesced. If allowed to be a few days undisturbed, the entire surface of the water became covered with a pellicle of the substance, which spread also over the stones and shells of the bottom and the sides of the vessel. It could be lifted in impalpable laminæ on sheets of paper. I found it difficult to keep it within bounds, and impossible to get quite rid of it, till after some months, I lost it by the accidental breaking of the vessel. Under the microscope this proved to be an *Oscillaria*, the filaments creeping and twining with the peculiar vermicular movements of the genus."[1]

WATER NET.

Certainly one of the most popularly interesting of fresh-water algæ is the water-net, not only on account of its singular appearance and comparative rarity, but also for some of the features in its life-history. It is well known that the water-net will rather suddenly make its appearance in some quiet water, where perhaps it was never known before, and be found there for two or three years, when it will as suddenly vanish, and the place will know it no more. It was found, not many years ago, in a small pond, less than half an acre in extent, in such profusion that no water could be taken out which did not contain it. Within three years most diligent search

[1] "Romance of Natural History," by P. H. Gosse, vol. ii. p. 104. (1861.)

did not reveal a single specimen, and none in the surrounding pools or the neighbouring ponds. About midsummer is its most flourishing period, afterwards diminishing, until with the autumn it disappears. In form it is usually oval or elliptical in outline, at first entire and un-broken, but soon torn, on account of its fragile nature, consisting of a delicate network of green threads, forming hexagonal meshes (Fig. 24). The whole net may be from two to six inches in length, and about an inch in dia-meter, distended only whilst floating. To view them properly they should be trans-ferred from the pond to a glass jar, in which the filmy nets will float suspended, and form most beautiful objects, as long as they last. The size of the meshes vary with age, being small when young, and larger, up to a quarter of an inch, when old. Examined closely, the whole net will be found composed of short slender rods or cells, equal in length to one side of the mesh. At the angles the ends of the cells are united, so as to form the mesh, and each cell is practically a single individual, so that the entire net is a colony of cells.

These cells do not communicate with each other;

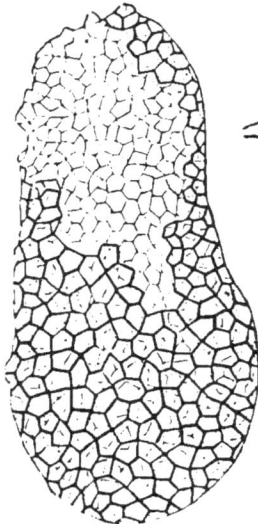

FIG. 24.—Water Net, *Hydrodictyon.*

they do not increase by cell division, but they can and do enlarge themselves as they progress towards maturity, each as a separate and independent individual. The cells are cylindrical, like rods, usually a little thickened at each end, containing an inner cylinder of green chlorophyl, floating in a watery fluid, and containing starch granules. The number of meshes, or rods, contained within an entire colony, or net, is just the same when full-grown as in an infantile state. The cell wall is double, the outer being a very thin membrane, and the inner more mucilaginous. The chlorophyl layer encloses the starch granules, which increase in number with the age of the cell, being at first only one or two, but finally several hundreds.

There is no sexual reproduction known, but the continuance of the species is provided for in two ways; that is, by the large gonidia, and the small gonidia, each with a separate history. The large gonidia are not so very much larger than the small gonidia, but they are concerned in the direct formation of the young net. The former are from seven thousand upwards in a single cell, the latter from thirty thousand. Braun says that "if a single fully developed net, or even fragments of nets, are placed in a shallow saucer of water, we may almost certainly reckon upon finding fully formed young nets in some of the old cells, on the next, or at all events on the second morning, and these in cells which exhibited no alteration whatever. If we wish to see the origin of these young nets we must not lose the earliest hours of the morning, for the tremulous movement commencing after the formation is complete, and the final union of the gonidia into a net, takes place shortly after sunrise, in the middle of summer between four and five in the morning, at the end of summer between six and eight o'clock ; and only in the dull days of autumn, in which however the formation of new nets rarely occurs, sometimes as late as ten o'clock in the morning. The gradual changes which take place in the preparation for fructification must be observed by night, but it sometimes happens that the particular cells do not advance far enough in the dissolving preparation in the course of the night to enable the influence of the morning's light to complete the formation. Such cells remain all day without any perceptible alteration, completing the preparatory stages in the second night and maturing the gonidia on the second morning."

The stages through which the cell contents pass in their differentiation into gonidia would be a little tedious to narrate in detail. Suffice it that the starch grains dissolve and disappear, the green becomes opaque, the plates of chlorophyl separate, the angles are rounded, a lens-shaped form is assumed, and finally a movement in the gonidia takes place and the large gonidia become united into a small net, which union begins even in the last period of the movement, so that gonidia already united in groups are seen still jerking and pulling backwards and forwards. When they have come to complete rest, and united into a new net, the roundish gonidia com- mence to elongate, and the first starch grain makes its appearance. The increase in size advances with astonishing rapidity. The cells of the net, under favourable circumstances, increase to six hundred times the length, and two hundred and forty thousand times the volume in a few weeks. Meanwhile the mother-cell becomes more and more gelatinous and ultimately melts away, leaving the young water-net free and exposed, ready to commence life on its own account. Thus much then, briefly, for the direct production of young water-nets by the large gonidia of a single mother-cell.

As for the other method of reproduction by the small gonidia, it takes place in cells indistinguishable in form, size, and appearance from those which produce the large gonidia. The gonidia themselves may be distinguished, not only by their smaller size, but by a longer shape, a small red vesicle, and four long cilia. They swarm out from the ruptured mother-cell, move about very actively, often for three hours, and then come to rest. The mother-cell, which contains at length the small gonidia, does not expand uniformly, but forms a bulging enlargement, at one or other part where the cuticle is torn, which bursts and lets out the microgonidia in a dense swarm, moving about most actively. The membrane of the cell, emptied of its contents, does not dissolve away at once, but remains for a long time unchanged as an empty coat. The swarming out always takes place rather later in the day than the formation of the nets, being in summer usually between seven and nine, and in autumn between ten and two o'clock. This swarming lasts for several hours, the active condition being often observed late in the afternoon, sometimes up to five o'clock in the evening. The swarming, or formation of nets, would seem to depend much on external circumstances. Some days only the large gonidia seem to be at work in the formation of young nets; on other days, and especially when the weather is dull and rough, the swarmers are produced in unusual numbers, and then there is no net formation.

When the water-nets disappear in the autumn and seem to leave no trace behind, it is some months before they reappear in the spring, and this interval is occupied, according to Pringsheim, by the "swarmers," or, as he calls them, the "chronispores," undergoing a period of rest or, more accurately, of hybernation. After moving about in the water actively for a few hours, they lose their cilia and come to rest; then they acquire a thicker outer cellulose wall and pass into the quiescent stage, in which state they are capable of existing in water for a long time.

After a period of not less than three months they recommence active life, and for some time after this there is little manifest change, except an increase of size. When they have acquired a diameter equal to one-fortieth of a millimetre the endochrome divides into several portions, the outer layers of the old cell wall give way, and the inner layers protrude like a sac, into which the contents pass and soon acquire all the well-known characters of true zoöspores. From two to five of these are developed from each "chronispore" or "resting spore." They are ovate, with two cilia, and, upon their escape from the sac, move above actively for a few minutes. After a little time they become motionless, lose their cilia, and develop into many-angled cells, with the angles prolonged into hornlike appendages. Under favourable circumstances, within a few days, the bright-green endochrome begins to undergo change, and is soon differentiated into gonidia, which unite within the parent-cell and form a miniature net, in the same manner as the large gonidia have been seen to combine in the original water-net cell; and, in due time, this young net is set free by the dissolution of the wall of the parent-cell. Thus a new individual is formed from and within the resting cell in the spring, in the same manner as the young nets are produced by the large gonidia in the cells of the mature nets in the autumn. And thus the cycle is complete.

GREEN VOLVOX.

"A thing of beauty is a joy for ever," and such is the little organism so well known to all lovers of pond-life as *Volvox globator* (Fig. 25). Quiet pools and clear ponds in woods and on commons often swarm with them; ever turning, rolling, floating in perpetual motion. Little green spheres, sometimes no larger than the head of a good-sized pin, seldom so ample as a small pea, must have their beauties revealed under a microscope, and hence, to the multitude, their very existence is unknown. Seen under favourable circumstances they are found to be spherical in shape, covered apparently by the most delicate network, as of the finest cobweb, with little green points scattered all over the surface, and yet so translucent that neighbouring objects can be seen through them. Other and smaller globes usually are seen to occupy the interior, revolving on their own account. Such are the objects which, notwithstanding their movements, are little green plants; but which at one time, on account of their motion, were believed to be animals. The whole true and particular account, such as recorded in scientific

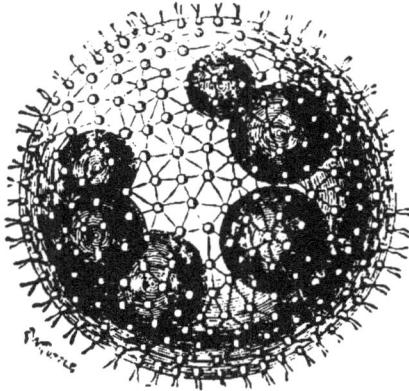

Fig. 25.—*Volvox globator.*

books, of their entire life-history, however interesting to the scientist, would be too prolix and technical for the general reader; but there are some of the most prominent features which may be touched upon with advantage.

Looking down upon one of these little spheres, which have been likened to crystal globes, we become conscious that the little green points, or dots, are arranged with mathematical accuracy, so that each one is in the centre of six others, placed at equal distances around it. These green points are the gonidia, having each the power, under favourable conditions, of becoming a perfect volvox. Each of these gonidia is almost pear-shaped, with the narrow end outwards, and furnished with a pair of movable hairs. Within the interior each has a red or brown eye-spot, and in all essentials agrees with

the common form of zoöspore found in so many of
the algæ, but, in this case, fixed at the periphery of
a revolving matrix in a combined family. By the
movement of the whole series of pairs of hairs, or
cilia, with which the gonidia are furnished, the little
spheres are kept in motion. By means of accurate
adjustment it will be observed that the gonidia are
connected with each other, telegraphically, by means
of two or three very delicate filaments, like the finest
of spider's web, so delicate that they can only be
distinguished with difficulty.

A certain number of these green points, or gonidia,
are privileged to produce a group of young individual
globes within the parent sphere, usually from four
to nine. Sometimes a daughter volvox, whilst still
enclosed within the parent sphere, will have in its
interior distinct evidence of the commencement of
its own daughter spheres. The process of development of the selected gonidia into young plants is as
follows: The gonidium enlarges, and becomes flattened and discoid, pressed to the inner surface of the
hyaline sphere, surrounded by eight ordinary zoöspores. Gradually the disc becomes oval with a
constriction across the centre. It then divides in
the opposite direction, so that it consists of four
segments. After this it protrudes more into the
inner cavity, and differentiation proceeds rapidly,
whilst a proper external membrane is developed.
Finally a green sphere, containing a vast number of
closely packed green granules, enclosed within a
proper envelope, is evolved. As this sphere enlarges
the granules separate from each other, leaving clear

spaces between, and become in fact the young
gonidia of the new individual. Several of these
young spheres appear almost simultaneously within
the parent, so that they are often closely packed ;
and yet each individual gonidium soon exhibits its
own pair of movable hairs. In this state the young
spheres await the rupture of the parent sphere to
escape into the water and commence a separate existence. The opening in the wall of the old volvox
is usually smaller than the diameter of the young
globes, so that the latter are compressed in their
exit, but at once resume the globular form. After
a little time, the parent, having accomplished its
mission, decays and dissolves away.

The process, as above described, is maintained
throughout the summer, and is practically a sort of
vegetative reproduction by division without any
intervention of sexuality. It is, in fact, an asexual
reproduction, which may be repeated many times ;
but at length is replaced, in the autumn, by a true
sexual process. In this we recognize the phenomenon of an alternation of generations ; several
generations are produced asexually by division, then
a sexual generation, ending in a hybernating oöspore,
which in turn produces another series of asexual
generations.

Some authors recognize two varieties, or even two
species, of *Volvox*, in one of which the male and
female elements are found on different spheres, in
the other the male and female organs are found on
the same sphere.

The female cells, when they are present, in the

autumn, are very similar to the ordinary neuter cells, only they are more numerous. The contents assume a dark green colour, at first having a frothy appearance from the formation of vacuoles, but ultimately become packed with protoplasm, but never undergo any process of subdivision. They acquire a flask-shape, with the narrow end to the circumference of the sphere. When ready for impregnation a spherical form is assumed, enveloped in a gelatinous membrane. The male cells at first more resemble the neuter cells, and when they have attained definite dimensions they begin to divide ; but they are of a lighter colour through containing less chlorophyl. They are ultimately resolved into a bundle of cells, each consisting of an elongated body, in which the green colour becomes reddish yellow, with a long and colourless beak, furnished at its base with two long movable hairs, and possessing also an eye-spot. The whole antheridium is enclosed within a gelatinous envelope, each of the contained bodies being a spermatozoid. These are fully developed just at the time that the oögonia, or female cells, are matured. By the breaking up of the antheridium the enclosed spermatozoids are set free and float actively within the gelatinous envelope, through which they speedily find their way into the cavity of the parent sphere. When entirely free they soon cluster around the female cells, or oögonia, and some of them penetrate the gelatinous envelope, and coalesce with the contents of the oögonium, which now becomes fertilized and secretes a new cell wall. This is at first smooth but afterwards covered with conical prominences, which gives it a stellate appearance. Simultaneously with this the green contents of the oögonia change to orange red, of an oily nature, and the oöspore becomes a resting spore. The mother-colony breaks up, whilst the thick-coated reddish oöspores sink to the bottom to hibernate, and this is the termination of another generation. This determines the question as to what becomes of the volvox during the winter, and we must follow the oöspores through their period of rest to ascertain how they reappear in the spring.

In the spring the resting spore gradually awakens to new life, and after some minor preparations the outer coating is ruptured, whilst the contents enclosed in an inner membrane protrude. These contents are soon observed to divide gradually into from two to sixteen, or more, small cells, which become again of a bright green colour, each one bearing at one extremity two movable hairs. These cells are gradually driven further apart from each other, by the interposition of gelatine, when, as the outer membrane disappears, the active hairs come into full play, and the young volvox is fairly launched into life. "The resting spores of volvox, therefore, germinate in water, and each of them produces a single colony by a process of segmentation, identical with that which gives rise to a daughter-colony, at the expense of a cell of the mother-colony."

With this initial step the cycle of non-sexual generations again commence, and is continued through the summer, until, at its close, resting spores are again formed by means of sexual developments, so as to be able to undergo the vicissitudes of winter.

"But who hath praise enough? nay, who hath any?
None can express thy works, but he that knows them;
And none can know thy works, which are so many,
And so complete, but only he who owes them.
All things that are, though they have several ways,
Yet in their being join with one advice
To honour thee! and so I give thee praise."

GREEN LAKE.

The name of Glas-lough, or "green lake," has been applied, from time immemorial, to a lake in the county of Monaghan, in Ireland, on account of the hue of its waters, the cause of which was investigated by Dr. Drummond in 1837, with the result that it was caused by the presence of minute vegetation. He says that "when a little of the water is lifted in the hand it seems perfectly transparent, and it appears equally clear at the edges of the lake, in a depth of not more than a few inches, and there the pebbles at the bottom show perfectly distinct, without any intermediate cloud to obscure them. But at a depth of two feet the bottom is undistinguishable, and the water presents a sort of feculent opacity, accompanied with a dull, dirty, greenish hue. On lifting some of this in a glass it seems at first sight quite transparent, but, on holding it up to the light, innumerable minute flocculi are seen floating through every part of it, and producing a mottled cloudiness throughout the whole."[1]

The microscope revealed that the minute flocculi, in suspension, were the very fine filaments of a species of fresh-water alga, belonging to the genus *Oscillaria*,

[1] "On a New Oscillatoria," J. L. Drummond, M.D., in *Annals of Natural History*, March, 1838.

which appears to have been detected nowhere else, and was called, from its colour, *Oscillaria ærugescens*. At first the plant was only found diffused through the water, but after a time a wet ditch was discovered running from the lake into an adjoining field, and in this ditch large masses, several inches in thickness, and above eighteen inches in length, swimming on the surface. These were evidently produced by an agglomeration of the filaments which had floated out of the lake, and matted together on the surface. The masses thus formed were tough and slippery, so that they could not be lifted out on the end of a stick. The surface when dried became of a bright verdigris hue, but where immersed, of a dull opaque green.

It was discovered, and confirmed by the inhabitants around the lake, that when the water has stood at rest in a vessel for a night or two, a green scum spreads over its surface, which is skimmed off before the water is used for domestic purposes. This confirms the tendency to aggregation in the filaments, and accounts for the masses found in the ditch. Similar masses are sometimes found floating at the sides of the lake, or cast on the shore, but the times and seasons at which this took place could not be ascertained.

Although the same county abounds in lakes, the phenomenon above described is peculiar to Glaslough, and in this, from all accounts, the green colour is evident throughout the year. The filaments, in the glomerated masses, were sometimes many inches in length, running parallel ; but the free filaments were shorter, and often broken.   Sometimes the oscillating motion was very vivid, but at other times it could scarcely be observed.   No investigation could throw any light upon the prevalence of this particular alga in this one lake.   There was nothing in its surroundings or the vegetation of the banks which would afford any clue to the mystery, and the story of the "green lake" remains a romance of low life.

Subsequently Dr. Dickie announced the appearance of the same alga, under similar circumstances near Aberdeen.   He says, " I have found the same species particularly abundant in a small and shallow artificial lake, in sheets of great extent at the bottom.   I have not observed it ' broken into innumerable fragments, and suspended like cloudy flocculi in the water.'   It sometimes, however, becomes detached from the bottom, and forms large masses on the surface." [1]   It is not by any means an unusual condition for the filaments of an *Oscillaria* to become separated and diffused through the water.   We may not know the precise condition under which this dispersion takes place, but it is a matter of experience, not only that *Oscillaria* will adhere together in large matted sheets of considerable extent, but that it will also under other conditions be severed and suspended.   There is still so much which we do *not* know in relation to the *Oscillariæ*, that what we really *do* know seems to be only a small proportion of what is unknown.   Why and how do the filaments oscillate?   How is each species propagated and continued?   Are there any reproductive organs, and of what character?   Or is

[1] *Annals of Natural History*, January, 1849, p. 20.

reproduction the result of a continued fission ? At present there is silence !

Speculations have not been wanting, but none with any great claim to plausibility. We will submit one which was proposed fifty years ago.

Dr. Unger, in the year 1840, expressed his views as to the nature of *Oscillatoria*, in which he enunciated his belief in their animal nature. "I did not propose," he says, "to enter into a comparative examination of the Oscillatoriæ, but to combat a system, according to which the forms at present known ought to be necessarily referred to some vegetable genus, composed of elements very heterogeneous. When Agardh, speaking of some Oscillatoria which move with the greatest ease, states that they have an articulated head which they move after the manner of a beak, he certainly by this points to an animal nature. The characters assigned by Agardh to the *Oscillatoria animalis* of Carlsbad are far more striking : according to his expressions, it does not oscillate ; it has not the pendulum-like movement ; but it crawls like a worm, and turns itself in every direction. It is also able to move itself freely in the water, differing thus from the others, which are only able to do this when they rest on the common substratum. It moves the head, which is linguiform, as the mollusca move their tentacula ; in a word, animal movement cannot be denied them. Moreover, if we refer to his own opinion, which he expresses when describing the Oscillatoria of Carlsbad, that their characters depend chiefly on their mode of life, we shall be so much the more induced to consider these productions rather as animal than vegetable. It is only to be regretted that we know so little of the history of the life and development of these beings, for we refuse to characterize some forms as animal on the sole ground that we do not find in them that movement which is the property of animal nature." [1] We must leave these speculations as they were, and proceed with our record, for manifestly they bring us no nearer to a solution.

Another instance of the abundant production of a minute alga in a lake, so as to impart a characteristic colour, is recorded by Mr. W. Thompson. " Late in the autumn of 1837," he says, " I observed patches of a singular bluish-green scum at the edge of Ballydrain Lake, a few miles from Belfast, but when I returned soon afterwards it had disappeared. On visiting the lake in July, 1838, I found that the whole body of water was tinged with a dull faint glaucous hue. On going out in a boat to ascertain the cause, I saw that the water was everywhere filled with extremely minute particles, which might be compared to motes in a sunbeam. To the unassisted eye they seemed as delicate as the finest human hair, and of a spiral form. Around the boat their motion was not very rapid, but those on the surface moved in an opposite direction from the particles beneath, and the latter the more quickly. On inquiry from some relatives, who reside on the borders of the lake, I learned that the appearance had been observed only for the last four or five years, and for about three months in each year. One of my friends had looked

---

[1] *Annals and Magazine of Natural History*, Nov., 1839, p. 214.

upon its approach with dread, as it interfered so much with his angling that during the period of its continuance this sport had to be abandoned. Eels, pike, and perch, especially the latter, are abundant in the lake, but when the water is clouded by the plant, the diminution in the number of perch taken is said to be not less than about one to fifty—the difference is attributed to the fish not being able to see the bait. Five days afterwards I found the entire lake tinged with this plant, but unequally so. In some parts where the water was two feet deep, the bottom could be seen ; in others it was invisible at one-fourth of this depth ; at the leeward and windward sides there was but little difference, except that at the latter it was occasionally observed to give a pale green tinge to the surface, where the water circling gently congregated it together, and threw it thence in a broken cloudy form for a moment, when it was again dispersed. From eminences at some little distance the green tinge of the water is most conspicuous ; in some places the colour is of a pale dull green, in others greenish brown, thus imparting the dull dead aspect of a Dutch canal.

"Twelve days afterwards the lake was found to be more densely coloured than before. The day being perfectly calm, the surface of the water was covered to some extent, where the depth appeared to be about five or six feet, with an alga of a pale but rich green hue. Ten days elapsed, and, on visiting the lake, the water had lost some degree of its opacity, and looked clearer. Instead of the beautiful appearance which the surface formerly presented, there was in some places merely a little scum, which, excepting its very pale greenish tinge, resembled precisely the appearance remaining on the surface of water in which ice has been dissolved. Towards the edge of the lake there were in some places gelatinous tufts of a pale blue colour, in one place crowded together in a mass which covered an area a few yards in extent. The portion nearest the edge had, apparently from decay, become ferruginous, and strongly tinged with rust colour the paper on which it was placed, but with the greatest pains I could hardly obtain a trace of the blue colour. The masses, both blue and ferruginous, were very slippery to the touch, about an inch in thickness, and of considerable consistence, and, on being lifted out of the water in a wire-gauze net, remained there without diminution by dripping off, or otherwise ; their weight, too, was great. When brought near they had somewhat of the offensive smell of water in which flax has been steeped, and at a short distance from one part of the lake this disagreeable odour was sensibly perceived. About seventy days after the first visit the algæ had entirely disappeared from the lake, the water throughout its depth as well as at the surface being clear and pure. The conclusions arrived at were, that the alga undergoes no change whatever, either in size or otherwise, from its first appearance as a colouring matter, until about three months afterwards, when decomposition ensues, and it is utterly dissolved. The specimens obtained of the alga were invariably of similar breadth, and rarely presented more than four spiral turns, and about one-fiftieth of an inch in length, at first of a dark green colour, when ascending to the surface singly of a pale green, and when in a mass of a pale blue, becoming ferruginous in decay. The species was determined to be a variety of *Anabæna flos-aquæ*, mixed in some places with *Aphanizomenon.*"[1]

Professor Allman[2] appears to have met with the same alga as that of the Ballydrain Lake, in the Grand Canal Dock at Dublin—at any rate, specifically the same, with some varietal modifications. It was in October, 1842, that he observed a substance, of a pea-green colour, abundant in the water of the docks. This substance was unequally distributed, being in some places collected in large quantity, while in others the water was quite free from it. It consisted of flocculent, unattached masses, varying much in size and occupying very different depths, some floating upon the surface, while others were observed suspended in the water, and might be traced downwards till the depth alone concealed them from the sight. The general appearance of these masses might remind one of certain substances in the act of precipitation, or of the curd of milk when diffused through the uncoagulable part of the fluid. In some places the green matter had been left by the retiring water upon the stones of the margin, and, here drying, had assumed a beautiful bluish-green or verdigris colour, without lustre. On visiting the docks at different periods, it was observed that the substance

---

[1] Detailed account in *Annals and Magazine of Natural History* for April, 1840, p. 75.

[2] "On a New Genus of Algæ," by G. J. Allman, in *Annals and Magazine of Natural History*, March, 1843, p. 161.

appeared sometimes in but very small quantity, while on other occasions it was to be seen much more abundantly. This appearance and disappearance seeming to be independent of the direct rays of the sun, but probably the result of meterological influences.

Under the microscope it was found to consist of exceedingly minute, simple, moniliform threads, with the globules composing them of uniform diameter, and the threads themselves variously but elegantly curved and grouped together, without order in a gelatinous matrix. Hence the conclusion that the substance was an alga, and very near to *Anabœna*, similar to the alga of Ballydrain Lake, which Professor Allman called *Trichormus spiralis*, and the present one *Trichormus incurvus*, but both now referred to *Anabœna flos-aquœ*.

It is remarked that Bory Saint-Vincent was so impressed with a belief in their animal nature that he removed the genus *Anabœna* from the vegetable kingdom. The peculiar motion of reptation which he describes them as possessing, and which he compares to the crawling of worms, would appear to have been the chief grounds on which he assumed their animality. No such motion was detected in the species of Ballydrain Lake, or the Grand Canal Docks. When a large mass of the latter was placed in a limited quantity of water, decomposition soon set in, the green colour becoming duller, and finally assumed a dirty, ferruginous hue. A disagreeable odour was at the same time exhaled, but this odour was altogether different from that of decomposing animal matter, and possessed a purely vegetable character.

RED SEA.

"The sea, then, makes a twofold indentation in the land upon these coasts, under the name of *rubrum*, or 'red,' given to it by our countrymen ; while the Greeks have called it Erythrum, from King Erythras, or, according to some writers, from its red colour, which they think is produced by the reflection of the sun's rays ; others, again, are of opinion that it arises from the sand and the complexion of the soil, others from some peculiarity in the nature of the water." Thus wrote Pliny[1] of the Red Sea a very long time ago, and since that time many persons have puzzled themselves, and others, to account for the redness of colour, attributed to the waters of the Red Sea. Perhaps the best summary of the past, and most satisfactory determination of fact, were contained in Montagne's memoir on the colouration of the water of the Red Sea, in 1844,[2] wherein the colour is attributed to an alga, allied to *Oscillaria*, and known by the name of *Trichodesmium erythræum*. The conclusions which are arrived at in this memoir, are—

1. That the name of the Erythræan Sea, given first to the Sea of Oman and to the Arabian Gulf by Herodotus, afterwards by the later Greek authors to all the seas which bathe the coasts of Arabia,

[1] "Natural History," bk. vi. ch. 28.
[2] "Sur la Coloration des Eaux de la Mer Rouge," *Ann. des Sci. Nat.*, 1844, p. 332.

probably owes its origin to the very remarkable phenomenon of the colouring of its waters.

2. That this phenomenon, observed for the first time in 1823 by M. Ehrenberg, in the bay of Tor only, then again seen twenty years later by M. Dupont, but in truly gigantic dimensions, is owing to the presence of a microscopic alga, *sui generis*, floating at the surface of the sea, and even less remarkable for its beautiful red colour than for its prodigious fecundity.

3. That the reddening of the waters of the Lake of Morat by an *Oscillaria*, which M. de Candolle has described, has the nearest relation to that of the Arabian Gulf, although the two plants are generically very distinct.

4. That, as we may well suppose, according to the accounts of navigators, who mention striking instances of the red colouring of the sea, these curious phenomena, though not observed till quite recently, have nevertheless without doubt always existed.

5. That this unusual colouring of seas is not exclusively caused, as Péron and some others seem to think, perhaps as being chiefly zoologists, by the presence of mollusca and microscopic animalcules, but that it is often also due to the reproduction, perhaps periodical, and always very prolific, of some inferior alga, and in particular of the species of the singular genus *Trichodesmium*.

6. That the phenomenon in question, although generally confined between the tropics, is however not limited to the Red Sea, nor indeed to the Gulf of Oman ; but that, being much more general, it is found

in other seas: for example, in the Atlantic and Pacific Oceans, as appears in the "Journal of Researches," by Darwin, and from unpublished documents of Dr. Hinds, from which the following is extracted.

"Dr. Hinds, who sailed in the ship *Sulphur*, sent to explore the western coasts of North America, first observed on the 11th of February, 1836, near the Abrolhos Islands, the same alga, doubtless, which Darwin saw at the same date. This alga was again seen many days running. Some specimens of it having been brought to Dr. Hinds, he perceived that a penetrating odour escaped from it, which had before been thought to come from the ship ; this odour much resembled that which exhales from damp hay. In April, 1837, the *Sulphur* being at anchor at Libertad, near St. Salvador, in the Pacific, Dr. Hinds again saw the same algæ. A land breeze drove it for three days in very thick masses about the ship. The sea exhibited the same aspect as at the Abrolhos Islands, but the smell was still more penetrating and disagreeable ; it caused in a great many persons an irritation of the conjunctiva, followed by an abundant secretion of tears. Dr. Hinds himself experienced it. The alga in question constitutes a distinct species of *Trichodesmium* (*T. Hindsii*). It differs from that of the Red Sea, both in dimensions and smell."[1]

Darwin's account is a brief one, for he says, "When not far distant from the Abrolhos Islets, my attention was called to a reddish-brown appearance in the sea. The whole surface of the water, as it

[1] *Comptes Rendus*, July 15, 1844; *Annals and Mag. Nat. Hist.* (Sept., 1844), vol. xiv. p. 225.

appeared under a weak lens, seemed as if covered by chopped bits of hay, with their ends jagged. These are minute cylindrical confervæ, in bundles or rafts of from twenty to sixty in each. Berkeley informs me that they are the same species with that found over large spaces in the Red Sea, and whence the name of Red Sea is derived. Their numbers must be infinite : the ship passed through several bands of them, one of which was about ten yards wide, and, judging from the mud-like colour of the water, at least two and a half miles long. In almost every long voyage some account is given of these confervæ. They appear especially common in the sea near Australia ; and off Cape Leeuwin I found an allied, but smaller, and apparently different species. Captain Cook, in his third voyage, remarks that the sailors gave to this appearance the name of 'sea-sawdust.'

Macdonald found the same, or a similar alga, off the coasts of Australia, and in Moreton Bay, amongst the Polynesian Islands, and on two separate occasions off the Loyalty Group, in nearly the same geographical position. He says, 'When the filaments are first removed from the water they may be observed adhering side by side in little bundles or fasciculi ; and, besides the colouring matter, the little cells, or at least the intervals between the septa, contain globules of air, which sufficiently account for their buoyancy ; and, moreover, in this respect, although their abiding-place is the open ocean, their habit can scarcely be regarded as very different from that of those species which flourish in damp localties exposed to the atmosphere. The filaments are all very short compared with their diameter, with rounded extremities ; and when immersed some little time in fluid so that the contained air-bubbles make their escape or are taken up, the pale colouring matter appears to fill the cells completely, and a central portion, a little darker than the rest, may be distinctly perceived in each compartment, intersected by a very delicate tranverse partition." [1]

RED SNOW.

Coloured snow-storms were recorded as long ago as the sixth century, and a shower of red hail was mentioned by Humboldt as occurring at Palermo. It is believed that De Saussure first noticed in Europe the red snow, in 1760, on Mount Breven in Switzerland, and subsequently so frequently amongst the Alps, that he was surprised that it should have escaped notice by all previous travellers. Ramond found it in the Pyrenees, and Sommerfeldt in Norway. In 1818 an Italian journal contains an account of the fall of red snow in the Italian Alps and on the Apennines. In March, 1808, the whole country around Cardoce, Belluno, and Feltri was covered in one night to the depth of nearly ten inches, it is said, with a rose-coloured snow. A pure white snow fell before and afterwards, so that the coloured snow formed an intermediate stratum. A like phenomenon occurred at the same time on the mountains of Veltelin, Brescia, Krain, and Tyrol. A similar one occurred at Tolmezzo, in the Friaul, between the 5th and 6th

[1] *Proceedings of the Royal Society*, Feb. 26, 1857 ; *Annals of Natural History* (May, 1857), vol. xix. p. 431.

of March, 1803; and a more remarkable one still in the night between the 14th and 15th of March, 1813 in Calabria, Abruzzo, Tuscany, and Bologna, consequently along the whole chain of the Apennines.[1]

Captain Ross saw mountains in Baffin's Bay which were covered by red snow, eight miles long. The snow was found to penetrate in some places to a depth of ten or twelve feet, and seemed to have existed long in the same state. Darwin, in his narrative of the voyage of the *Beagle*, relates that, when travelling in the Andes, he saw a mountain covered with red snow. At a meeting of the San Francisco Microscopical Society, Dr. Harkness presented a bottle of red snow which he gathered on the Wasatch Mountains. It was found about ten thousand feet above the sea level, and, when fresh, had the appearance of being drenched with blood, as though some huge animal had been slaughtered.

The early opinion prevalent was that the red snow fell from the sky, that it invariably fell during the night, and consequently no one ever saw it fall. At one time it was doubted whether it was a lichen, a fungus, or even an infusorial animal. Finally it was determined to be a fresh-water alga, to which the present name is applied of *Chlamydococcus nivalis*. Like other algæ, moisture seems essential to its production, and hitherto this plant has not been found in places where it was debarred from this pabulum at some period of its growth. But once formed, it seems to possess the power of remaining stationary, and, perhaps, of reviving after an unlimited period.

[1] J. C. Agardh, " Memoir on the Red Snow."

In the Arctic region it was discovered on snow, on rocks, on decayed mosses, and on the bare soil. In Scotland its locality is curious. The island of Lismore, in which it is found, is very low, ten miles in length by only one or two in breadth, and resting on a limestone rock of a slate-blue colour. "It occurs," says Carmichael, "in abundance on the borders of the lakes of Lismore, spreading over the decayed reeds, leaves, etc., at the water's edge, but in greater perfection on the calcareous rocks within the reach of occasional inundation; and, what is rather remarkable, it seems to thrive equally well, whether immersed or exposed to the dry atmosphere. It is to be found, more or less, at all seasons of the year."

Dr. Greville examined some of the Lismore specimens, and some from the Arctic regions, with the following results: "I had them immersed in water for a period of three weeks, but did not succeed in tracing any appearance that was not developed equally well in the course of a few hours. In every instance I found no difficulty in detecting a gelatinous substratum, various in thickness (sometimes exceeding the diameter of the globules), colourless, diffuse, without any defined border. Upon this gelatine rests a vast number of minute globules, the colour of fine garnets, exactly spherical, nearly opaque, yet very brilliant, for the most part nearly equal in size; the smaller ones generally surrounded with a pellucid limb like the capsules of *Ceramium*, and this limb gradually becoming less as the globules enlarge, at last entirely disappearing. In the full-sized globules a favourable light shows the existence of internal

granules, which make the surface to appear reticu-
lated. When mature they burst, and the internal
granules escape, to the number of six or eight or
more, and the membrane only of the globule is left
behind, buoyant and colourless. The granules are
globose, and escape from the globules one by one,
adhering together, though I never could observe the
least voluntary motion among any of these bodies."

In the year 1878, Brun noticed on the sacred
mountain near the city of Ouessin, in Morocco, a
so-called "rain of blood," which he found to result
from a quantity of minute shining flakes, which
adhered closely to the rocks and presented an extraordinary resemblance to drops of blood. These
were found to be a young and undeveloped condition
of *Chlamydococcus*, mixed with organic remains and
fine sand. He suggested that they had been brought
by a strong south-west wind from the Sahara, where
the *Chlamydococcus* is assumed to be extremely abundant. There is a record extant of the occurrence of
red snow in Hertfordshire, in 1881, but the account
there given is rather mixed, and even the inference
that some condition of *Euglœna* caused the red appearance serves in no way to elucidate the mystery.[1]

Martius, who twice accompanied French expedi-
tions to Spitzbergen, writes of a green field of snow
which was seen on the coast of Spitzbergen, in July,
1838. The surface of the snow was white, but a few
centimetres below it was deeply coloured, as if it
had been sprinkled with a decoction of spinach. In

[1] M. C. Cooke, "Introduction to Fresh Water Algæ," pp. 170-172.
(1890.)

another instance, Martius found this green substance scattered like dust over the surface of a snow-field, the greater part of which was covered with an immense mass of *Chlamydococcus nivalis*, below the surface, and on the edges of the field the snow was also coloured green. From many observations, Martius came to the conclusion that the red globules of the green snow are identical with those of the red snow, and that the green snow and the red are one and the same plant, only in different stages of development. It is remarkable that this conclusion, which was arrived at half a century ago, coincides with the accepted opinion of to-day. The great difficulty which presented itself to most of the old observers was the presence of a condition when the sphærules took the form of animalcules, or were active zoöspores. Those who regarded them, with Ehrenberg, to be animalcules, called them *Enchelis.* "M. Agardh, in 1823, saw in red snow that the globules, generally considered as vegetables, sometimes pass into animalcules ; and the behaviour of the *Enchelides* in the passive and active state is one reason why so many philosophers have spoken of a metamorphosis of infusoria into plants."[1] Ehrenberg relates of the *Enchelides* that several are still quite green, while others appear spotted, half red and half green ; and this might perhaps be taken as the best proof that these so differently coloured Infusoria belong to one and the same species. In order to understand, and see the bearing of these

[1] "Red and Green Snow," in *Ann. and Mag. Nat. Hist.*, June, 1841.

observations, it is necessary that the life-history of *Chlamydococcus* should be borne in mind, from which it will be evident that all these organisms, instead of being distinct, are but separate conditions of the one species, which is *Chlamydococcus nivalis* (Fig. 26).

It is considered not only possible, but very probable, that the red-snow plant is not specifically distinct from the organism long known as *Protococcus pluvialis*, and hence that the life-history of the one is the same at that of the other. The details we possess of the latter are more explicit than of the former, and from these we may present an outline of their peculiar characteristics. Normally full-grown cells of this organism, which sometimes re-

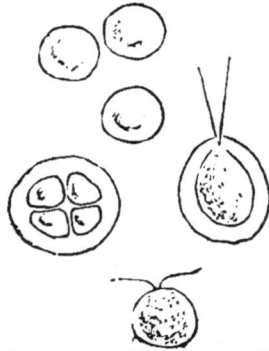

Fig. 26.—Red Snow, successive stages of development.

sembles a plant and sometimes an animal, are glo-bose, with a thick tough cell membrane, and opaque granular contents, sometimes of a brown and some-times of a bright red colour. Within the cell contents lie hidden starch granules and a nucleus. There also appears to exist in the centre of the cell a large nuclear vesicle, so covered by the rest of the cell contents as to be indistinctly perceived. When these resting globular cells are placed in water, they give birth to four gonidia-like swarming cells. Even before the commencement of division of the contents, a change begins in the colour of the parent cell, the red colour retreating from the circumference, and a yellow (or greenish) border forming round the deep-red central mass. The young swarmers, for a short time after they issue, have only a narrow yellow rim round a dark red middle. During the two or three days' period of movement and growth of these swarming cells—in which they attain to four times the original size, changing their ovate to a pear-shaped form,— important changes take place in their contents. The red colour becomes more concentrated into the middle of the cell, so that a sharply defined red nucleus is formed, with a lighter space in the interior, corresponding to the nuclear vesicle, around which the red colouring matter forms a covering more or less complete. The rest of the cell contents have become a brilliant green, and in them the starch granules, with smaller green granules, are visible. The ciliated point of the cell, often drawn out like a beak, is colourless. This first moving generation is succeeded by a number of similar active genera- tions, populating the water for some weeks, often giving it a bright green colour, until, at length, uni- versal rest recommences, and the cells sink to the bottom, or attach themselves elsewhere. The transi- tion from one active generation to another passes through a resting generation of extremely short dura- tion. The full-grown swarming cells finally come to rest within their wide envelope, and almost simul- taneously divide into two cells, which, without becoming active, divide again into two cells. Thus, within the parent envelope four daughter-cells are produced, which begin to move soon after they are completely formed, and, rupturing the thin envelope, make their escape. The whole of this process is very

rapid, being completed in one night and the next
morning. The second active generation thus formed
resembles the first, except that the active cells are
green from the first, and have a smaller red nucleus
in the interior. The subsequent active generations
bear a general resemblance to the preceding, but
with many modifications. For example, the full-
grown swarm cells not unfrequently assume strange
two-lobed or four-lobed shapes, beginning to divide
before they come to rest, or sometimes a transverse
constriction and bisection of the cell takes place.
The formation of vacuoles is a pretty constant
phenomenon in the later active generations, and
there may be many of them excentrically placed,
with the red nucleus retaining its central position,
or a single central vacuole, causing a lateral displace-
ment of the red nucleus. The red nucleus often
becomes very small in the last generations, so that
it resembles the red corpuscle which Ehrenberg
called "eye-spot."

Not uncommonly the red colour wholly disappears.
In the later stages of the "cycle of generations" the
formation of microgonidia takes place. Many individuals, instead of producing four daughter-cells,
undergo further division, so as to produce a brood
of sixteen or thirty-two minute cells, which at first
form a mulberry-like body, at length separating to
commence a very active swarming inside the parent
cell, which ends in a rupture of the envelope, and
the dispersion of the little swarmers. These are of
a more elongated shape than the large swarmers, of
a yellowish or dirty yellowish green colour, with

reddish ciliated points. They do not increase in size, never become coated with a loose membrane, and have no further power of propagation. Most of them die after they have settled to rest, and dis-solve away; others turn into little red globules, but it is doubtful if they grow to the normal size.

The large swarmers of the last active generation, when their growth is completed, and they have attained the stage of rest, instead of dividing again, remained undivided, assume a perfectly globular form, and, in a few days, become covered with a thick, closely applied, cell membrane, while the earlier loose membrane wholly disappears. The contents, which when the resting commenced were green with a little red nucleus, or entirely green, gradually become red again, passing from green through many tints of brown, or brilliant golden-green and golden-brown, into red. These globose thick-coated cells (the same as those with which we began) behave like spores, and pass into a condition of perfect rest. They exhibit no growth, and, after the membrane has attained its proper thickness and the contents their red colour, no perceptible further alteration.

In order to complete the main features of these alternating generations, it must be noted that, in addition to the active generations (macrogonidia and microgonidia) and the concluding generation, passing into resting spores, there are other generations, which are the proper representatives of the vegetative de-velopment. These are generations endowed with quiet and slow vegetative growth, multiplying by

pure **vegetative** division, without any swarming movement. It depends upon external conditions whether
the resting cells at once give rise to new active
generations, or to a series of quietly vegetating generations of cells. The formation and multiplication of
these vegetative generations takes place by the
division of the cell contents, either by simple division, the first generation being transitory, or by
double halving. But the newly formed cells do not
slip out, as do the swarmers, from the mother-enve-
lope, they remain in the same place and position.
The membrane of the mother-cell seems to become
softened, expands, and becomes gradually drawn out
to nothing, and at length vanishes, the daughter-cells
in the mean time acquiring a tolerably thick, closely
applied, cell membrane of their own. The division
is repeated many times in this way, and as the cells
all remain in intimate contact, small families, and at
length large conglomerations of cells, are produced.
Ordinarily the colour is light brown. The above,
therefore, briefly represents the main features of the
life-history of this small organism, which has often
presented itself as a puzzle for the curious, but
presents many features of peculiar interest.[1]

GORY DEW.

Another of the minute fresh-water algæ, which is
known to botanists by the name of *Porphyridium
cruentum*, has in times past caused considerable

[1] For further details, consult " British Fresh Water Algæ," by M. C.
Cooke, p. 51 (1882) ; Braun's " Rejuvenescence," p. 206 ; Cohn "On
Protococcus Pluvialis" (Ray Society, 1853).

alarm amongst the uneducated by its habit of sudden appearance in blood-like patches upon damp walls. Drayton says that "in the plain near Hastings, where the Norman William, after his victory, found King Harold slain, he built Battle Abbey, which at last, as divers other monasteries, grew to a town enough populous. Thereabout is a place which, after rain, always looks red, which some have attributed to a very bloody sweat of the earth, as crying to Heaven for vengeance of so great a slaughter." This substance is nevertheless quite a natural production, and is common enough on the lower part of damp walls, in cellars, dairies and outhouses, on the ground, gravel walks, and hard-trodden paths, and is most conspicuous after rain. It forms broad patches of a deep blood-red or purple colour, with a shining surface, as if blood or red wine had been poured upon the ground. Examined by the microscope, it is found to consist of an agglomeration of minute globose cells, filled with granular matter. We have met with it, commonly, on the stone or tiled floor of a conservatory, in large irregular port-wine patches, which make their appearance sometimes very suddenly, and are calculated to alarm superstitious and uneducated people. On one occasion we remember that a damp corner in a country church was pointed out to us, with some reserve, as something weird and mysterious, on account of patches like dark blood-stains appearing on the stones at certain periods of the year. Dark-olive patches, having some resemblance, except in colour, are found at the base of walls, but these are caused by a common species of *Oscillaria*.

ANIMAL DESMIDS?

The species of *Closterium* are amongst the largest and most beautiful of the Desmids (Fig. 27). They are often distinctly visible to the naked eye, of a

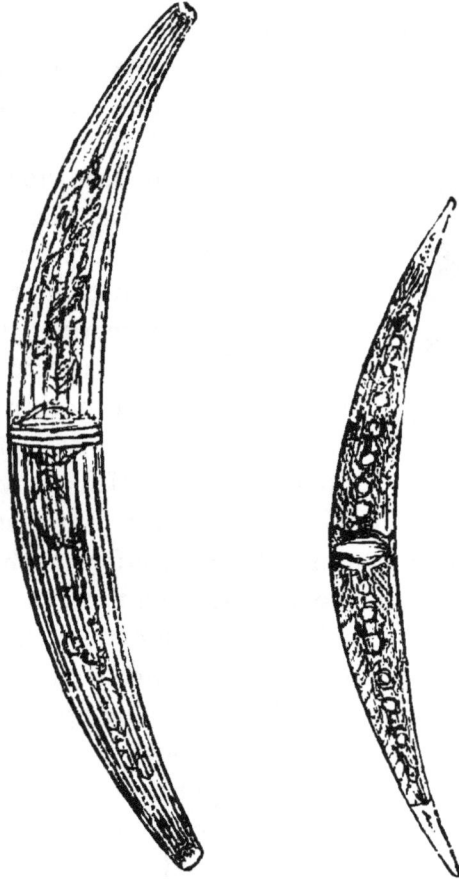

FIG. 27.—*Closterium Striolatum; Closterium Leiblcinii.*

beautiful green colour when living, and flourish in the clear water of ditches, streams, and bogpools. In form they are always elongated, more or less curved, sometimes lunate, and attenuated towards each end. A paler transverse band crosses the centre of the frond, and the ends are usually occupied by a clear space containing active granules (Fig. 28). They vary in size according to the species. At the present time they are universally accepted as algæ, reproducing themselves by zygospores, resulting from conjugation. Ehrenberg, however, regarded them as belonging to the animal kingdom, for the following reasons, viz. "Because they enjoy voluntary motion, they have apertures at their extremities, they have projecting permanent organs near the apertures, which are constantly in motion, and they increase by horizontal spontaneous division." Dr. Meyen, who was of the opposite opinion, contends that "their structure is exactly similar to that of the Confervæ, their formation of fruit and the development of the fruit also like that of the Confervæ. The occurrence of starch in the interior, with which they are frequently nearly filled, is a striking proof of their being plants ; they have no feet, for what Ehrenberg regarded as such are molecules, having a spontaneous motion, the function of which it is difficult to determine." In 1839 Mr. Dalrymple advocated their animal nature in a communication to the Microscopical Society of London, giving as his reasons the following summary :—

1. That while *Closterium* has a circulation of mole-

FIG. 28.—*Closterium lunula*, with active granules at the end. The arrows indicate the direction of surface circulation.

cules greatly resembling those of plants, it has also
a definite organ unknown in the vegetable world, in
which the active molecules appear to enjoy an independent motion, and the parietes of which appear
capable of contracting upon its contents.

2. That the green gelatinous body is contained in
a membranous envelope, which, while it is elastic,
contracts also upon the action of certain reagents,
whose effects cannot be considered purely chemical.

3. The comparison of the supposed ova with cyto-
blasts and cells of plants precludes the possibility of
our considering them as the latter, while the appearance of a vitelline nucleus, transparent but molecular
fluid, a chorion or shell determines them as animal
ova. It was shown to be impossible that these eggs
had been deposited in the empty shell by other
infusoria, or that they were the produce of some
entozoön.

4. That while it was impossible to determine
whether the vague motions of Closterium were voluntary or not, yet the idea the author had formed of a
suctorial apparatus forbade his classing it with plants.

Lastly. In no instance had the action of iodine
produced its ordinary effects upon starch or vegetable
matter, by colouring it violet or blue, although Meyen
asserts it did in his trials.

Hence the author concluded that Closterium must
still be retained as an infusory animal.[1]

Seven years later M. Eckhard contended still
for the animal nature of these Desmids. He said
that "the grounds for their being of animal nature

[1] *Annals of Natural History* (August, 1840), vol. v. p. 415.

are derived partly from their motion and partly from their organization." On the leaves of *Ceratophyllum* he observed the manner in which several *Closteria* adhered elegantly by one extremity ; in about a quarter or half an hour many of them were situated in the same manner upon a higher part of the leaf ; not a single animalcule was found on the side of the leaf nor adherent longitudinally to it. They had evidently moved during the above time from the lower to the upper part of the leaf. If we observe their motions under the microscope they are not so rapid as those of many other polygastric infusoria ; but the motion is always evidently animal. They swim, especially in summer, in the most varied directions, and he had frequently seen them swim against the current when the water on the object holder was flowing towards one side, whilst fragments of plants, various kinds of *Spirogyra* and *Oscillaria*, were carried away. It is difficult here to discover anything but animal motion ; to explain this, however, by electricity, as Turpin attempted, is unnatural, and not less absurd than that of the muscular fibre by the same natural agent, by Strauss. But the relations of the organization of the *Closteria* are likewise in favour of their animal nature.[1] He then alludes to form, terminal vesicles, the supposed "feet" of Ehrenberg, the rows of granules, etc., and concludes, "All these are not plant-like ; and if the carapace of *Closteria* should prove to be of a horny nature, they would be removed from the vegetable kingdom with still greater certainty."

[1] *Annals of Natural History* (Jan., 1847), vol. xviii. p. 434

PANSPERMISM.

Fifty years ago a very reckless heresy was prevalent in Germany, which afterwards gained some ground with superficial thinkers in this country, that, under varied and changed conditions of heat, moisture, etc., the same germ was capable of producing widely different objects. Many books were written in which these views were inculcated, and supported by the vaguest of arguments, and inconsequent or unsatisfactory observations. Here and there were a few earnest men, whose judgment was misguided, but whose opinions were entitled to respect, since doubtless to them their conclusions seemed to be logical, and loyal to their religious belief. The extent to which these views were carried may be predicated from the following observation, made by Riessek of Vienna,[1] who claimed to have "succeeded in making pollen grains germinate in the parenchym of leaves and stems, not merely of the mother-plant, but also on those of others belonging to different natural orders ; that they produced fungi laden with spores, and that these spores when placed in water produced confervoid plants filled with chlorophyl, and copulating with one another ; that he observed also the metamorphosis of the pollen-cells into animals of Ehrenberg's genus *Astasia*, and that the contents of the pollen-cells also produced plants and animals. From the smaller particles originated Bacteria, Vibrios, and Confervæ ; and from the larger, green globular monads." He

[1] *Botanische Zeitung*, July 19, 1844.

M

professed also to have observed the transformation of the chlorophyl of flowering plants into Confervæ and Infusoria.

The celebrated Kutzing did not proceed so far as this, but he professed some most extraordinary views. In a prize essay his object was to show that the lower forms of algæ were capable of being changed into more highly organized species, or even into species belonging to different families, and classes, of the higher cellular plants. Subsequently, in another treatise, he extended his observations to Infusoria, and believed that he had observed their transmutation into algæ.[1] We are not prepared to go over the evidence and report the conclusions, otherwise than in the general manner indicated above, but we may record the impressions and observations which were made at the time upon the inferences drawn by him from his experiments. "Firstly, the observations cannot be considered conclusive, apart from all prejudice either way, till a certain number of bodies, ascertained to be precisely of the same nature, be isolated, and the changes of these observed with every possible precaution to avoid error. At present, it seems that there is not by any means sufficient proof that the objects in question really arise from germs of the same nature. Secondly, there appears too often in treatises of this description to be great indistinctness as to the notion of what a species really is. We know that in the course of develop-ment higher bodies go through a vast variety of

[1] F. T. Kutzing (Nordhausen), "Uber die Verwandlung der Infu-sorien in niedere Algenformen." (1844.)

phases, which resemble very closely true substantial species which have arrived at their full development ; but we are not therefore to suppose that in passing through their phases the production has really consisted of such a number of real species. In the Agardhian sense this may be true enough, for when he pronounces the vessels and cells of flowering plants to be algæ his meaning appears to be, however strongly he expresses himself, merely that they are representatives of algæ, and resemble them in structure."

Finally, the real difficulty of the case does not depend on the question as to the difference of animal and vegetable life. These evidently, in certain parts of the creation, are so intimately combined, that it is quite impossible to say where the one ceases and the other begins ; and there is really no reason why we should be incredulous as to the possibility of the same object appearing at one time endowed more especially with animal, and at another with vegetable life. Late observations on the reproductive bodies of some algæ show that their motion is produced by vibratile cilia, exactly in the same way as in certain animals. But it is exceedingly difficult to imagine the transformation of one real species into another. The same species may assume a vast variety of forms, according to varying circumstances, and it is highly instructive to observe these changes ; but, that the same spore should under different circumstances be capable of producing beings of an almost entirely different nature, each capable of reproducing its species, is a matter which ought not to be admitted generally, without the strictest proof. Observations made with care on isolated individuals, and not on a common mass, which can scarcely be otherwise than more or less heterogeneous, could not fail to be instructive, and might lead to results which, if they did not confirm the views so commonly entertained in Germany, would have had an influence on science, which it is difficult at present to appreciate.[1]

AGUE PLANT.

Early in 1862 intermittent fever began to show itself in the malarial districts of the Ohio and Mississippi Rivers. The disease rapidly increased during the months of July and August, till it had invaded nearly every family on ague levels. Dr. J. H. Salisbury, having previously been engaged in a series of experiments on certain plant diseases, came to the conclusion that some close investigation might lead to the discovery of the source of intermittent fever in the exhalations of malarial districts. The result of these experiments he recorded in 1866.[2] The investigations were commenced by a microscopical examination of the expectorations of those labouring under intermittent fever, and who resided upon ague levels, and were exposed during the evening, night, and morning to the cool, heavy, damp exhalations and vapours rising from stagnant pools, swamps, and humid low grounds; in short, those who were constantly im-

[1] *Annals and Magazine of Natural History* (Dec., 1844), xiv. p. 433.
[2] "On the Cause of Intermittent and Remittent Fevers, etc.," J. H. Salisbury, M.D., in *American Journal of Medical Sciences* (1866), vol. li. p. 51.

mersed in a malarial atmosphere, and where every
one was more or less affected with symptoms of
miasmatic poisoning.

The first salivary secretions and mucous expectorations of the morning were those used. In these
secretions, amongst other organisms, he found as the
only constant bodies, uniformly in all cases, and
usually in great abundance, minute oblong cells,
either single or aggregated, consisting of a single
nucleus, surrounded by a smooth cell wall, with a
highly clear, apparently empty, space between the
outer cell wall and the nucleus. Their peculiar appearance satisfied him at the outset that they were
not fungoid, but cells of an algal type, resembling
strongly those of the Palmellæ.

After satisfying himself that these minute cells
were the only forms found that could be relied upon
as constantly present on malarial levels, and not
present above them, his next step was to trace their
source, if possible, and their character. The details
need not be entered upon, but suffice it to say that
rectangular plates of glass were suspended about one
foot above the surface of stagnant pools and marshy
grounds that were partially submerged. On both
surfaces of the plates the oblong cells were found in
considerable numbers, and these experiments were
repeated for many nights, in varying localities, with
the same results.

Incidentally it is remarked that, in walking over
the swampy grounds in order to suspend the plates,
the doctor noticed a peculiar dry feverish sensation
always produced in the throat and fauces, extending

often to the pulmonary mucous surface, and that his expectoration was, after returning, uniformly filled with the minute oblong cells above described. Other medical men, who accompanied him on subsequent occasions, experienced the same results. "Numerous other persons," he says, "who visited with me ague grounds, were invariably affected with the same train of symptoms." The only constant bodies found in the expectoration of those affected with the above local symptoms produced by walking over ague grounds, and in the expectoration of those immersed in the night emanations of malarial levels, were the minute palmelloid cells previously described. The source of these cells was found to be the palmelloid plants growing in such profusion on the drying soil of ague lands during the prevalence of intermittents. It is thence inferred that the minute cell emanations from these low vegetable organisms are capable of exciting local fever in the mucous surfaces with which they come in immediate contact; and, further, that there is strong presumptive evidence, from what has been previously determined, that, by repeated and continued exposure to them they may cause general fever, of either an intermittent or remittent type. This was Dr. Salisbury's contention; and then he adduces cases in proof, one of which may be cited, although adduced subsequently. He says, "After exhibiting, about the 1st of November, a large pan of soil covered with this vegetation to the class in one of my lectures, I placed it under the working-table in Dr. House's office. It was loosely covered with a newspaper and forgotten. In a few days the doctor began to have pains in the back and limbs. These symptoms were soon followed by a well-marked paroxysm of ague. As soon as this occurred the pan of plants came to my mind, and was removed."

We may pass on to the more precise details which are given of these ague-plants, and it seems to be rather singular that, instead of being referred to well-known genera or species, they should all be placed in three *new* genera, constructed specially to receive them. It may be added that, during a quarter of a century, these three new genera have failed to secure for themselves a place in systematic algology. One of these genera which includes half a dozen species, which are characterized chiefly by colour, is termed *Gemiasma*, including "plants having the appearance of cells, each consisting of a thin outside wall, enclosing an inside cell filled with minute spores, either single or aggregated, multiplying by duplicative segmentation within a parent membrane, and also developed from spores." All these genera, it is stated, have spores of a similar structure: the spores are mostly oval, or more or less oblong, and have double walls.

"The species are many, all of which have heretofore been regarded as innocuous. There is strong evidence for believing, however, that the minute species that are developed in such abundance in the above-mentioned localities, and the spores of which become elevated and suspended in such multitudes in the heavy humid night exhalations of ague districts, are decidedly poisonous to the epithelial surfaces with which they come in contact, and are the true source of intermittent and remittent fevers."

We have quoted the words in which Dr. Salisbury sums up his conclusions as to the source and origin of ague, but we fear that the whole account must now be regarded as a pretty "fairy tale," and that the true source of intermittent fevers must be sought in another direction. There is no ground for supposing that the learned doctor acted otherwise than in perfect good faith, and truthfully described all he saw and experienced; the initial facts may have been perfectly true, and yet his inferences be proved false and untenable. The little romance vanishes before the strong light which has been thrown upon the Bacteriological theory of epidemic disease.

In 1872 Dr. Bartlett, of Chicago, sent over to this country some mud containing plants of that which he conceived to be the ague-plant of Dr. Salisbury. He says, " I sought for the plants described by him in the ague bottom of the Mississippi River, opposite Keokuk, Iowa." [1] " By placing the cake of earth, sent you, in a plate, and adding water enough to make it of about the consistence of potter's clay, and keeping it at a temperature above 60°, you will find a fresh crop of the plant to develop, and you will thus have an opportunity of studying them. Should you allow them to flourish and remain uncovered in your room, you might have the satisfaction of demonstrating the 'cause of ague.' This plant was first found, so far as I know, by Dr. J. P. Safford, of Keokuk, who was kind enough to search for me while I visited a fever patient. In the locality of their growth they are to be seen in myriads, and near

---

[1] *Grevillea* (1872), vol. i. p. 95.

them, even on elevations of over one hundred feet, everybody had the ague. The course of this disease seemed *pari passu* with that of the plant." Shortly after this period Dr. Bartlett read a communication before the Chicago Society of Physicians and Surgeons, in which he fully described what he be-

lieved to be the ague-plant ;[1] whilst, from the details and descriptions, American botanists arrived at the conclusion that the plant described was no other than the well-known fresh-water alga, called *Botrydium granulatum.* This had already been described and figured from British specimens, irrespective of any suspicion that it was associated with the ague.[2] It is unnecessary to enter upon any details of the structure here, since they are available in books on the subject,[3] especially as the charges have not been

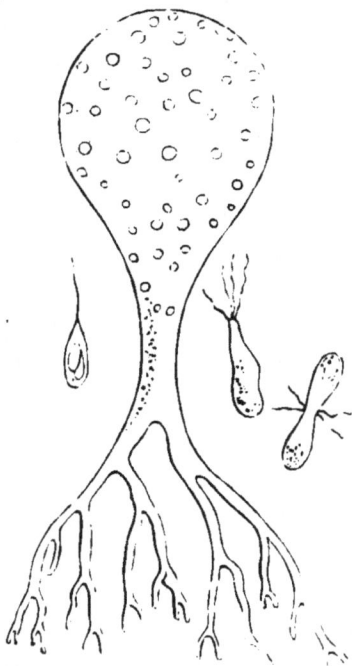

FIG. 29.—*Botrydium granulatum,* with zoöspores.

established. The mud sent over by Dr. Bartlett was, however, submitted for examination to a most accomplished authority; and in 1874 Mr. William Archer confirmed the opinion that the "ague-plant" in

[1] *Grevillea* (1874), vol. ii. p. 142.
[2] *Ibid.* (1872), vol. i. p. 103, pl. 7.
[3] Cooke's "British Fresh Water Algæ," p. 114. (1884.)

question was no other than the harmless *Botrydium granulatum* (Fig. 29).[1] He says that, on reading over the description by Dr. Bartlett, "one sees how fairly it tallies with the known characters of the *Hydrogastrum* (otherwise *Botrydium*), but it is undoubtedly surprising how he, and the American observers of the Society referred to, failed to perceive the identity of the organism in question—one which finds a place in so many botanical text-books, both by figure and description, as well as on lecture diagrams, as a noteworthy example of a single-celled independent plant, and at the same time endowed with the power to become copiously ramified, so to speak, root, stem, and aerial portion combined in one cell only. I venture to think it hardly less surprising to find this seemingly so passive and inert little chlorophyllaceous alga, met with in suitable situations all over Europe, gravely *tried, and found guilty*, on so slender evidence, of being the atrocious 'cause of the ague.'"

The "ague-plant," in so far as *Botrydium granulatum* is concerned, may therefore be regarded as a romance of the past, but its history will remain as a lesson to those who, upon suspicion only, and without more than the slightest circumstantial evidence, would gravely bring a charge of complicity in disease against a harmless and innocent little organism. That this little alga favours such swampy localities as those in which ague usually abounds may be granted, but that by no means proves an association of the supposed cause and effect.

The more feasible interpretation of the probable

[1] *Grevillea* (1874), vol. ii. p. 166.

cause, or, at the least, the associate, of intermittent and malarial fevers, is the presence of a bacillus in the blood attacking the red corpuscles. This bacillus, which has been named "*Bacillus malariæ*, is abundantly found in the blood of patients during the period of attack, while during the period of acme which terminates the attack only spores are found. The same microscopic organism is found in all the malarious districts of the Roman Campagna, and it can be produced in artificial culture."[1] Richard writes that the multiplication of these bodies must be extremely rapid. For instance, in tertian fever they are not found in the intervals of the attacks (apyrexia). As the attack approaches they appear in increasing numbers, and their maximum corresponds with the beginning of the rise in temperature; from that moment they begin to perish, since the heat of the fever is fatal to them, and completely checks their development. This explains the intermittent character of the disease. They produce fever, the fever kills them and then subsides, when apyrexia occurs they multiply again, excite fever, and so on. Thus there is a successive series of auto-infection by the parasite itself, unless its development is arrested by sulphate of quinine.

DIATOMS, IF ANIMAL.

One of the most fascinating romances connected with low life was that which held sway some half a century ago, strengthened by the support of the venerable Ehrenberg, that the Diatoms were animals,

[1] Trouessart, "On Microbes, etc.," p. 182.

and not plants. The strongest evidence in support
of this theory was undoubtedly the peculiar spontaneous movements of which they were capable, such
movements, it was contended by Ehrenberg, being
produced by the protrusion of cilia or feet from the
extremities of the frustule. Subsequent observers
failed to detect these organs of locomotion. Undoubtedly the old definition of the attributes of
plants and animals had something to do with
creating a prejudice in favour of the animal nature
of diatoms ; this definition summarily disposed of
the question by pronouncing that animals were
organisms endowed with locomotion, and plants were
organisms without locomotion. Afterwards, when it
came to be demonstrated that some animals, like
sponges, had no locomotion, and many plants, especially algæ, were locomotive, it became much easier
to believe that the mere fact of locomotion did not
prove animality. Under these changed conditions
the *nature* of the motion in the disputed organisms
came under discussion, it being contended that the
movements were entirely different from those which
characterized other locomotive plants, were conducted by instinct or intelligence, and consequently
were of an animal nature. " If nature," writes
Humboldt, " had endowed us with microscopic
powers of sight, and if the integuments of plants
were transparent, the vegetable kingdom would by
no means present that aspect of immobility and
repose under which it appears to our senses." After
it came to be doubted whether any organs of locomotion existed, such as Ehrenberg described, we find

that writers such as Meneghini, who contended for the animal nature of the Diatomaceæ, qualified it by the supposition "that in a being whose nature, for other reasons, we believe to be animal, the movements may be effected by the admirable vital powers through organs which escape our sight by their minuteness."

In a letter from M. Petit, quoted by Montagne,[1] he says, "The motion of these *Naviculæ* is well worth attention. It is more or less decided according to the degree of their development. In their more or less rapid progress across the field of view, they appear to have a certain degree of consciousness, so as to avoid any obstacles with which they meet. They advance for the purpose of investigation; they try the obstacle with one of their extremities: but they appear to do this with a certain degree of precaution. It seems as though they smell at these obstacles, that they examine them, and try means of avoiding them. I may add that I am quite certain that the movements of these little creatures do not depend on currents arising from the evaporation of the fluid on the stage, or from any other physical cause, of which it is easy, with a little attention, to convince one's self. These movements are certainly self-dependent, for the creatures wander in different, and frequently opposite, directions; and they consist, not simply in an agitation without object, but seem to be directed by a sort of instinct. On carefully watching them, we see them turn round obstacles

[1] Montagne, "Sylloge Crypt.," p. 471 (1856). *Annals Nat. Hist.* (1856), xvii. p. 278.

when they cannot pass above or below them. Sometimes, when they are entangled in a mass of dead organic matter, they put it in motion by their struggles to extricate themselves. You may therefore consider as certain all that I tell you about the spontaneous motions of our *Navicula*, which I scarcely regard as a vegetable (Fig. 30)."

Professor Leidy observes, independently, and apparently with no hypothesis in his mind, that "he encountered a small species of *Navicula*, very active, in a rain pool, and found it could move grains of sand, as much as twenty-five times its own superficial area, and probably fifty times its own bulk and weight, or perhaps more." [1]

FIG. 30.—*Navicula rhomboides.*

The movements in some species are very peculiar, and those of *Bacillaria paradoxus* are well known, and often cited. Mr. Thwaites has thus described them : " When the filaments have been detached from the plants to which they adhere, a remarkable motion is seen to commence in them. The first indication of this consists in a slight movement of a terminal frustule, which begins to slide lengthwise over its contiguous frustule ; the second acts simultaneously in a similar manner with regard to the third, and so on throughout the whole filament, the same action having been going on at the same time at both ends of the filament, but in opposite directions. The central frustule thus appears to remain

[1] *Proceedings of Academy of Nat. Sciences, Philadelphia* (1875), p. 113.

stationary, or nearly so—while each of the others has moved with a rapidity increasing with its distance from the centre, its own rate of movement having been increased by the addition of that of the independent movement of each frustule between it and the central one. This lateral elongation of the filament continues until the point of contact between the contiguous frustules is reduced to a very small portion of their length, when the filament is again contracted by the frustules sliding back again, as it were, over each other ; and this changed direction of movement proceeding, the filament is again drawn out until the frustules are again only slightly in contact. The direction of the movement is then again reversed, and continues to alternate in opposite directions, the time occupied being generally about forty-five seconds. If a filament while in motion be forcibly divided, the uninjured frustules of each portion continue to move as before, proving that the filament is a compound structure, notwithstanding that its frustules move in unison. When the filament is elongated to its utmost extent, it is extremely rigid, and requires some comparatively considerable force to bend it, the whole filament moving out of the way of any obstacle rather than bending or separating at the joints. A higher temperature increases the rapidity of the movement."[1]

Without attempting any further description of these movements, which are well known to all those who have studied these organisms, we may observe that, although the hypothesis of ciliary motion has

[1] *Proceedings of the Linnæan Society*, vol. i. p. 311.

been abandoned, no entirely satisfactory explanation of the causes of motion has been promulgated. The general view is that the movements are owing to forces operating within the frustule, and probably connected with the endosmotic and exosmotic action of the cell. Professor Smith was of this opinion, and his work still ranks as a text-book of British Diatomaceæ; and he believed that the fluids concerned in these actions must enter and be emitted through the minute foramina at the extremities of the siliceous valves. It may easily be conceived that an exceedingly small quantity of water expelled through these minute apertures would be sufficient to produce movements in bodies of so little specific gravity.

The following is a summary of the arguments which have been adduced in favour of the animal nature of diatoms :—

1. The Diatomaceæ—many species at least—exhibit a peculiar spontaneous movement, which is produced by certain locomotive organs.

2. The greater part have in the middle of the lateral surface an opening about which certain round corpuscles are situate, which become coloured blue when placed in water containing indigo, like the stomach-cells of many Infusoria, and consequently may equally be regarded as stomachs.

3. The shells of many Diatomaceæ resemble in structure and conformation the calcareous shells of *Gasteropoda* and similar Molluscs.

4. The method of multiplication by self-division.

5. The complicated structure of the wall of the frustules, and the characters of the siliceous deposit.

6. The greater affinity in chemical composition of the contents (endochrome) with animal than with vegetable products.

To all these arguments a sufficient answer has been given—some are not proved, others are not pertinent, and all are relegated to the records of past delusions. In all systematic classifications of the present day the Diatomaceæ retain their position as conjugating algæ and, of course, as a portion of the vegetable kingdom.

METEORIC PAPER.

It is one of the romances of past history that in the year 1686 a shower of a substance called "meteoric paper" fell from the sky in Courland ; but its true character was not determined until 1838, when it was examined by Ehrenberg,[1] and these results have been summarized in the following manner. "On the January 31, 1687, a great mass of a paper-like black substance fell with a violent snow storm from the atmosphere near the village of Rauden,

FIG. 31.—*Pinnularia major*.

in Courland. It was seen to fall, and after dinner was found at places where the labourers at work had seen nothing similar before dinner. This

---

[1] C. G. Ehrenberg, "Ueber das im Jahre 1686 im Courland vom Himmel gefallene Meteorpapier, etc.," in *Berlin Bericht der Akad* (1838), pp. 45-58.

C. G. Ehrenberg, "Mikroscopische Analyse des curländischen Meteorpapiers von 1686, etc." (Berlin, 1839; folio.)

N

meteoric substance, described completely and figured in 1686 and 1688, was recently again considered by M. von Grotthus, after a chemical analysis, to be a meteoric mass ; but M. von Berzelius, who also analyzed it, could not discover the nickel said to be contained in it ; and Von Grotthus then revoked his opinion. It is mentioned in Chladni's work on meteors, and noticed as an aerophyte in Nees von Esenbeck's valuable appendix to R. Brown's " Botan. Schriften." " I [Ehrenberg] examined this substance, some of which is contained in the Berlin museum (also in Chladni's collection), microscopically. I found the whole to consist evidently of a compactly matted mass of *Conferva crispata*, traces of a *Nostoc*, and of about twenty-nine well-preserved species of Infusoria, of which three only are not mentioned in my large work on Infusoria, although they have since occurred, living, near Berlin ; moreover, of the case of *Daphnia pulex?* Of the twenty-nine species of Infusoria only eight have siliceous shields, the others are soft, or with membranous shields. Several of the most beautiful, exceedingly rare, *Bacillariæ* are frequent in it. These Infusoria have now been preserved one hundred and fifty-two years. The mass may have been raised by a storm from a Courland marsh and merely carried away ; but may also have come from a far distant district, as my brother, Carl Ehrenberg, has sent from Mexico forms still existing near Berlin. Seeds, leaves of trees, and other things of the kind scattered through the mass, were on the examination of larger portions, easily visible. The numerous native Infusoria, and the shells of the

Daphnia pulex, seem to speak thus much for the substance, that its original locality was not the atmosphere nor America, but most probably either East Prussia or Courland." [1]

The light thus thrown upon the "Meteor paper which fell from the sky" proves it to have been thin sheets of a compacted *Cladophora*, or fresh-water alga, intermixed with the siliceous valves of several species of diatoms (called Infusoria by Ehrenberg) and some other minute objects.

Another similar substance is often found, in these latter days, to which the name of "vegetable flannel" or "natural flannel" is applied. This is found cast up on meadows where streams have overflowed and deposited masses of confervæ upon the grass and rushes. This substance is felted in sheets not unlike flannel in appearance, and bleached to a yellowish white by exposure to the sun and air. When it is examined it is found to consist chiefly of interwoven filaments of some species of *Cladophora*, with which are often mixed the frustules of minute species of Diatomaceæ. The origin of the two substances appears to be the same, as also their composition; but this last has not assumed such a compact papery texture as the former, and, being thicker and heavier, is not so liable to being blown away to a distance, to the dismay of the peaceable inhabitants.

Dr. Hooker collected in Sikkim, in the dry bed of a stream, a curious white substance, like thick felt

---

[1] "On a Meteoric Paper which fell from the Sky, etc.," by Professor Ehrenberg, of Berlin, in *Annals of Natural History* (1839), vol. iii. p. 185.

formed of felspathic silt, no doubt the product of glacial streams and siliceous cells of Diatomaceæ. It much resembled, he says, "the fossil or meteoric paper of Germany, which is also formed of the lowest tribes of fresh-water plants."[1]

DUST SHOWERS.

Many accounts have appeared, from time to time, of dust showers, which puzzled the curious for many years, until the discovery was made by Ehrenberg that they were composed for the most part of microscopic organisms, of which the greater proportion were minute diatoms.   The number of showers which Ehrenberg records is 340, of which 81 dated before the Christian era, and 249 since.   Details of some of the more recent are given ; for instance, in the Atlantic, about five hundred miles from the coast of Africa, "the dust was collected by Mr. Darwin from the ship in which he was at the time.   The direction of the wind was from the African coast. The dust resembled volcano ashes, although evidently not of this origin, and about a sixth part of it was siliceous shells of fresh-water and land Infusoria (diatoms), and siliceous phytolites—eighteen species of the former, and as many of the latter.   Most of the forms are European, and none exclusively African. Among them was the South American species, *Himantidium papillo*, which occurs in Cayenne, and also a *Surirella*, probably from the same continent."   The conclusion follows, as Ehrenberg observes, that either the dust came in part from South America in the upper region of the atmosphere, or these two species are yet to be discovered elsewhere.

Other Atlantic showers were examined from collections made by Darwin between the years 1834 and 1838, part at San Jago, Cape de Verde, and part within two hundred and fifty miles of the land in the open sea. They afford thirty new forms to those of the above shower, and include also the same South American forms. In addition there are three species of *Eunotia*, which have been found only in Senegambia and Guiana, together with *Amphidiscus obtusus*, also South American. The only new species was *Eunotia longicornis*, which is very similar to a Hungarian fossil species. No species peculiarly African was found in the dust ; one occurs in the Isle of France.

Of a shower which fell at Malta, May 15, 1830, the dust was obtained by Darwin, from the purser of the *Revenge*. The wind at the time was east-south-east, and a similar fall of dust took place at the time in the bay of Palmas, in Sardinia. The number of species afforded was forty-three. Some of the species occur in Africa, yet there are no characteristic African forms. And although such showers, with the hot winds that attend them, are usually referred to the Sahara desert, they appear to be quite foreign to that region. Among the species *Synedra entomon* is a characteristic form from Chili. In general character the species are like those of the Cape Verde and other Atlantic showers.

In the sirocco dust of Genoa, May 16, 1846, Ehrenberg found forty-six species. The forms have much resemblance to those of the Malta and Atlantic showers. The colour is yellowish, or ochreous, from oxide of iron, and not grey like the true African dust, about one-sixth of the mass being organic. None of the species are characteristic African forms, and *Synedra entomon* is South American. Hence it is concluded that the showers of the Atlantic, of Malta, and of Genoa are in general alike in organic as well as inorganic constitution, and in the absence of characteristic African forms ; and this resemblance is the more surprising as the observations extend through the long period of sixteen years. They are also alike in the brownish-red colour of the dust.

The sirocco dust which fell at Lyons, October 17, 1846, afforded sixty-seven species, besides minute portions of plants, of which thirty-nine belonged to Diatomaceæ. In this shower the organic forms make up about one-eighth of the mass. In general character, including colour, there is a close resemblance to the products of the Atlantic showers, and the others above described. The species are nearly all of fresh-water or land origin, one-seventh only being marine species. There are two South American species, the *Eunotia pileus*, and *Himantidium zygodon*.

A storm of so-called red snow in Puster Valley, in the Tyrol, March 31, 1847, owed its colour to a coloured dust, much resembling that of the Atlantic, its tint being brownish red. It afforded, as obtained from two localities, sixty-six organic forms, one-third of which were diatoms. The large majority of the species are known as fresh-water and continental forms, only two species being marine. There is a

remarkable resemblance in the colour and character
of the dust to that of the Atlantic, Genoa, and Lyons,
and an identity in many of the species. Forty-six
species out of the sixty-six occur in the Sirocco and
Atlantic dust. Twelve species of Diatomaceæ are
common to the Atlantic showers and the Tyrolese
snows. This uniformity of character over regions so
widely separate, yet in nearly a common latitude or
zone, and in so many distinct examples through a
number of years, is most surprising.

Dust also fell in Italy in 1803, and Calabria in 1813.
The former of these showers is represented as coming
from the south-east. It afforded forty-nine species,
and that of Calabria sixty-four. Out of the forty-nine
species thirty-nine have been observed in the more
recent showers, and fifty-one of the sixty-four. These
showers, ten years apart, have twenty-eight species in
common. In both, nearly all the species are of fresh-
water or continental origin. •

Ehrenberg remarks that these showers appear to
prevail most within a zone extending from the part
of the Atlantic off the west coast of Middle and
North Africa, along in the direction of the Mediter-
ranean Sea, reaching a short distance north of this
sea, and continued into Asia between the Caspian
Sea and the Persian Gulf, perhaps to Turkestan,
Kaschgar, and China; and they seldom reach north
to Sweden and Russia. This zone, according to the
observations of Tuckey, has a breadth of eighteen
hundred miles in the North Torrid zone. The
reddish colour of the dust, as well as the organic
forms, show that the dust is not of African origin.

Moreover, the storm winds and sirocco are found to afford the same species of organisms.

Ehrenberg repeats the opinion that these phenomena are not to be traced to mineral material from the earth's surface, nor to revolving masses of dust material in space, nor to atmospheric currents simply; but to some general law connected with the earth's atmosphere, according to which there is a self-development within it of living organisms.

Tables are added by Ehrenberg to his work on this subject, containing an enumeration of the species and their occurrence in the different specimens examined.

Chladni has estimated that for the single dust shower of Lyons in 1846, the material that fell was full 7200 hundredweight. The Cape Verde shower had a breadth, according to Darwin, of more than 1600 miles, and according to Tuckey of 1800 miles, and extended 600 to 800 miles, or even 1000 miles from the African Coast. This gives an area of 960,000 to 1,280,000, or from 1,684,000 to 1,854,000 square miles.

The surface of Italy is about 90,000 square miles; that of Sicily 10,000 square miles, making together 100,000. A single dust shower, covering both countries, like that of 1803, to the extent of that of Lyons in 1846, would deposit 112,800 hundredweight of dust in a single day. With such facts before us, Ehrenberg asks, how many thousand millions of hundredweight of microscopic organisms have reached the earth since the period of Homer, the time of our earliest record of such events? He adds, " I cannot longer doubt that there are relations according to which living organisms may develop themselves in the atmosphere." He supposes it probable that the atmospheric dust-cloud region is of vast extent, and is above a height of 14,000 feet.[1]

Whilst accepting the facts determined by Ehrenberg, we are far from adopting his theory of a vast dust cloud in "which living organisms may be developed." This is undoubtedly a romance which has no substantial basis in reason or in fact. Against the assumption that these minute organisms have been swept up by the force of wind, or currents of air, held for some time suspended, and finally been precipitated in clouds, like 'showers, no very strong argument can be adduced. The evidence is too strong to be rejected, that showers have been from time to time precipitated, and that, in cases where samples of that dust could be obtained, and submitted to examination, it has been found to consist, in very large proportion, of organized matter, and much of that consisting of the frustules of minute Diatomaceæ.

FOSSIL DIATOMS.

None of the lower forms of vegetable life are more common or more widely diffused than the Diatomaceæ. These were at one time, on account of their spontaneous movements, called Bacillaria and Infusoria, and suspected to belong to the animal kingdom. More complete investigation, however, determined them to be algæ, having a silicious or flinty skeleton. Typically, two similar thin translucent plates of silex,

[1] "On the Infusoria, and other Microscopic Forms in Dust Showers," by C. G. Ehrenberg, in *Silliman's Journal* (1851), xi. p. 372.

variable in size and form, are superimposed, with a band surrounding the margin, so as to leave a space between them. When all that is soluble is dissolved away the flinty skeletons remain, and these may be preserved for ages in immense deposits, practically indestructible, to testify to their previous existence. When the older authors wrote of animalcules and Infusoria, they for the most part included the diatoms; the residue were chiefly foraminifera and radiolaria, but most of the silicious deposits consisted mainly of Diatomaceæ.

"The stratum of slate, fourteen feet thick, found at Bilin in Austria, was the first that was discovered to consist almost entirely of minute flinty shells. A cubic inch does not weigh quite half an ounce; and in this bulk it is estimated there are not less than forty thousand millions of individual organic remains! This slate, as well as the Tripoli, found in Africa, is ground to a powder and sold for polishing. The similarity of the formation of each is proved by the microscope; and their properties being the same, in commerce they both pass under the name of Tripoli. One merchant alone in Berlin disposes annually of many hundred tons in weight. The thickness of a single shell is about the sixth of a human hair, and its weight the hundred and eighty-seven millionth part of a grain. The well-known Turkey stone, so much used for the purpose of sharpening razors and tools; the rotten stone of commerce, a polishing material; and the pavement of the quadrangle of the Royal Exchange, are all composed of infusorial remains."[1]

[1] Jabez Hogg, "The Microscope," p. 431. (Sixth edition, 1867.)

We take into account the large number of num-
mulites and other distinctly animal shells, belonging
to dead foraminifera, which go to form masses of
limestone rock, and which are often referred to under
the vague title of infusorial remains. It is sufficient
for our purpose to demonstrate that immense beds,
many miles in extent, in different parts of the world,
are entirely composed of diatoms. The material best
known is that called cementstein, from the island of
Mors, a large island in Liimfjord. This fjord, the
most extensive in Jutland, runs from east to west,
connecting the North Sea with the Kattegat. The
cementstein from Mors resembles a dark-grey slate,
interspersed with white veins. The siliceous organisms
of which it is chiefly composed are held together by
a calcareous cement, and when submitted to the
action of acids are slowly disintegrated with efferves-
cence. A similar deposit occurs in Fuur; it is, however, more difficult to separate, and but slightly affected
by acid, and resembles the deposit known as "brown
coal." A third deposit is found in Nijkjobing, a small
town or village on the western side of Mors Island.
This deposit is of a greyish-white colour, still more
difficult to reduce than the preceding, strong acids not
affecting it in any appreciable degree; and only by
the assistance of caustic potash or soda can the
organisms of which it is composed be effectually
separated. The Mors deposit, with perhaps one
exception, is richer in bizarre and beautiful diatomaceous forms than any other hitherto discovered.[1]

[1] See "Diatomaceous Forms from Jutland," by F. Kitton, in
Quekett Microscopical Journal (1870), vol. i. p. 99.

The *bergh-mehl*, mountain meal, in Norway and Lapland has been found thirty feet in thickness; in Saxony twenty-eight feet thick; and it has also been discovered in Tuscany, Bohemia, Africa, Asia, the South Sea Islands, and South America; of this, almost the entire mass is composed of flinty skeletons of Diatomaceæ. That in Tuscany and Bohemia resembles pure magnesia, and consists entirely of a shell, called *Campilodiscus*, about the two-hundredth part of an inch in size.[1]

Darwin says that the coast beds of shells which run along hundreds of miles in Patagonia "are covered by others of a peculiar soft white stone, including much gypsum, and resembling chalk, but really of a pumiceous nature. It is highly remarkable, from being composed, to at least one-tenth part of its bulk, of Infusoria. Professor Ehrenberg has already ascertained in it thirty oceanic forms. This bed extends for five hundred miles along the coast, and probably for a considerably greater distance. At Port St. Julian its thickness is more than eight hundred feet."[2]

"At Richmond, in Virginia, United States, there is a flinty marl many miles in extent, and from twelve to twenty-five feet in thickness, almost wholly composed of the shells of marine animalcules (Diatomaceæ); for, in the slightest particles of it they are discoverable. On these myriads of skeletons are built the towns of Richmond and Petersburg. The species in these earths are chiefly *Naviculæ;* but the most

[1] Jabez Hogg, "The Microscope," p. 432.
[2] Darwin, *Journal of Researches*, p. 171. (1852.)

attractive, from the beauty of its form, is the *Coscino-discus*, or sieve-like disc, found alike near Cuxhaven, at the mouth of the Elbe, in the Baltic, near Wismar, in the guano, and the stomachs of oysters, scallops, and other shell-fish. Another large deposit is found at Andover, Connecticut ; and Ehrenberg states that similar beds occur by the river Amazon, and in great extent from Virginia to Labrador."[1]

The testimony of Sir Joseph Hooker may be adduced as to the formation of like deposits in recent times in the Southern Hemisphere. "The Phonolite stones of the Rhine and the Tripoli stone contain species identical with what are now contributing to form a sedimentary deposit (and perhaps at some future period a bed of rock) extending in one continuous stratum for four hundred measured miles. I allude to the shores of the Victoria barrier, along whose coasts the soundings examined were invariably charged with Diatomaceous remains, constituting a bank which stretches two hundred miles north from the base of the Victoria barrier, while the average depth of water above it is three hundred fathoms, or one thousand eight hundred feet."[2]

They abound also amongst the pancake ice of the South Pole, as far south as seventy-eight degrees, where they must occasionally be subject to a very low temperature. Though much too small to be discernible by the naked eye, they occur in such countless myriads as to stain the berg and pack ice, wherever they are washed by the swell of the sea ;

[1] Jabez Hogg, "The Microscope," p. 433. (1867.)
[2] Berkeley, "Introduction to Cryptogamic Botany," p. 127.

and when enclosed in the congealing surface of the

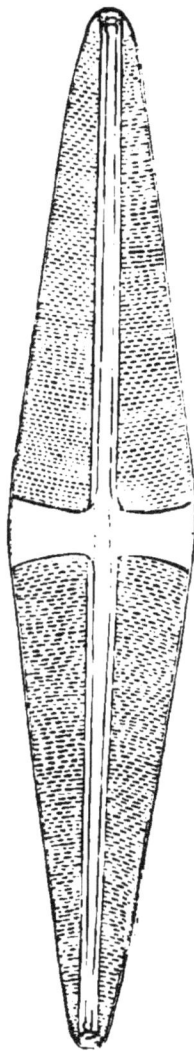

FIG. 32.—*Stauroneis Bayleyi*, from Monmouth Deposit. (*Science Gossip*.)

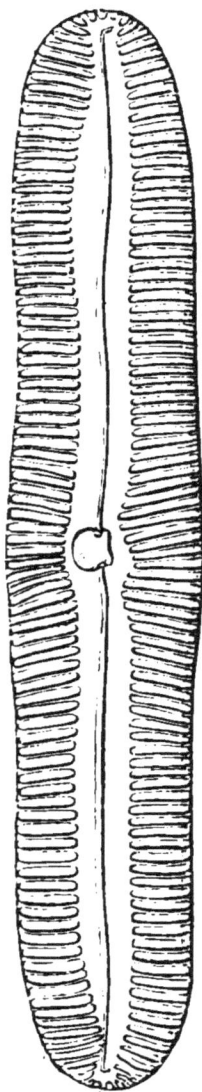

FIG. 33.—*Pinnularia gigas*, from Monmouth Deposit. (*Science Gossip*.)

water, they impart to the brash and pancake ice a pale ochraceous colour.

As representing some of the forms of the diatom valve, as found in deposits, we give figures of two of the larger forms from the Monmouth deposit, found in Maine, United States. They belong to the two genera of diatoms respectively known as *Stauroneis* (Fig. 32) and *Pinnularia* (Fig. 33), of which many species are known, both recent and fossilized.

Still more beautiful are some of the discoid forms

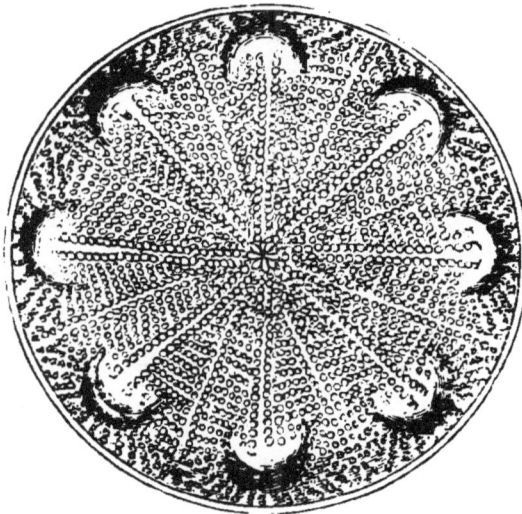

Fig. 34.—*Aulacodiscus Kittoni*, from Peruvian guano.

which are found in guano and elsewhere, and the quadrangular or star-like forms from similar sources. Of these the species of *Aulacodiscus* (Fig. 34) was obtained from Peruvian guano, and the *Amphitetras* (Fig. 35) from guano of Algoa Bay. These can give little idea of the variety of form and markings of the very numerous species already known. These valves or plates, being composed of silex or flint, are practically indestructible by fire, acids, or weather, although, being brittle as glass, they may easily be broken by force, and hence are often met with in fragments or in an imperfect condition. From these skeletons nothing could be determined of the structure and life

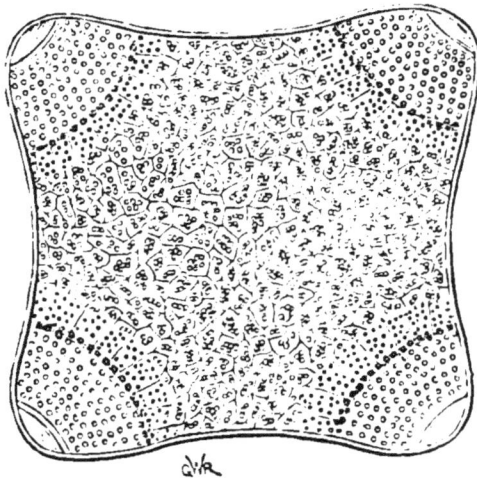

FIG. 33.—*Amphitetras ornata*, from Algoa Bay guano.

of the original organisms, but this can all be supplemented from the study of living plants, so that we can determine their place in nature, to which end we submit the following observations. They are essentially aquatic plants, or algæ, and, as one author observes, " I suppose there is scarcely a single piece of water anywhere which does not contain at least some individuals of the commoner species. They are to be found alike in the lake that crowns the mountain-top and the swamps and peat-beds which fill the lowest valley ; in the watercourse employed to irrigate the meadows; in the broad ocean, and the shallow puddle left by the overflowing of a ditch: the brackish water, where the tidal river meets the sea; salt-works and salt-pits; even inland lakes, which have a trace of salt in them,—each affords a rich variety of characteristic Diatomaceæ, varying according to the chemical quality of the water. They are to be frequently found also on rocks and masses of stone, damp from overhanging trees, or from the constant trickling of water."

The Diatomaceæ are now acknowledged to be true one-celled algæ, differing in many points from other algæ, but especially in the nature of the cell wall, which is composed of silex or flint, secreted by a thin membrane; also by their mode of increase by cell division; and by the nature of their cell contents, which contain a colouring matter not found in other algæ, and which is called *diatomine*.

"The structure is most easily discerned in the circular forms. In its earliest condition the cell has the appearance of a double-convex lens, but, as growth proceeds, the convex surfaces recede, and a band is formed by the sides, as of a pill-box, which gradually increases in breadth. If we take two pill-boxes, minus the lids, which we call *a* and *b*, the latter small enough to slide into the former, we have a fairly correct idea of the diatom cell. When this cell has attained its maximum development, the cell contents begin to divide, one portion receding to one end of the cell and the second to the other, and in a short time a new end is formed to *a*, and we have a cell in all respects like the first. The same occurs with *b*. We have now two perfect cells, one of which is smaller than the other, by the thickness of the band. In the free forms the cells usually separate, but in many genera the cells continue attached, thus forming a filament or chain. This process goes on until the productive energy is destroyed, but the number produced by the parent cell is probably very large. We have many times measured the thickness of this band, and never found it exceed the ten-thousandth of an inch in thickness.

"The shape of the diatom cell is very variable: they occur circular, oval, triangular, square, five to eight or ten-sided, boat-shaped, wandlike, sometimes curved in opposite directions, sometimes only in one. But the structure is always the same, viz. the two lids and the band connecting them. The former are usually termed *valves*, and the latter the connecting zone, or *cingulum*. It is from the form of, and the markings on, the valves that the generic and specific distinctions are principally constituted. These markings are generally of great beauty, and usually require the highest microscopic powers to show all their details, particularly those forms on which the markings appear as fine lines. One species in particular has lines so close that as many as ninety-four in the one-thousandth of an inch have been counted, and by certain arrangements of the light these lines are found not to be continuous, but composed of minute dots. The nature of these markings has always been a disputed question, some asserting that they are elevations on the surface, whilst others have been equally positive that they are minute cells in the siliceous wall (Figs. 36, 37). That the latter theory is true in some species has, we think, been clearly demonstrated in some of the large marine forms, but it is not so evident in the more delicate species.

The movements of the diatoms excited the wonder

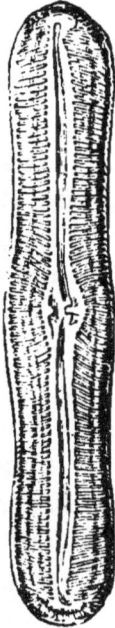

FIG. 36.—*Pinnularia major.*

FIG. 37.—*Stauroneis phenicenteron.*

and curiosity of the earliest observers, and was a great argument in favour of their animal nature. The means by which these movements are produced has not yet been explained. By some it was supposed to be caused by the extension and retraction of a foot or feet, and by others the expulsion of water from the ends of the cell. It is only among the elongated, or staff-shaped forms, that these movements are conspicuous, but that all non-parasitic forms possess the power of locomotion can be proved by placing a gathering containing them in the sunshine, when they will be found in a short time on the surface mud, appearing like a film of a golden-brown colour.[1]

The claims which the diatoms possess to special mention in a work devoted to what may be termed the "romance of low life in plants," are based on their immense number in nature, consequent on their indestructibility, their collection in vast beds or deposits of variable depth and extent; and the continuous formation of these deposits down to our own times; not to mention their beautiful and artistic sculpture, their flinty structure, and the peculiarity of their spontaneous movements.

The manner in which the deposits of these flinty valves have been formed is very simple. The living plants flourished in some lake or other expanse of water, and, as they died, the skeletons became disengaged, and, being heavier than water, fell to the bottom, forming a thin layer over the mud. Year after year, and century after century, the little plants continued to multiply, flourish, and die, thus contributing their skeletons to the stratum at the bottom, until at length, when the lake dried up, or its waters were diverted, the skeletons still remained, as an incontrovertible evidence of the existence of themselves, and the lake in which they flourished ages ago.

---

[1] F. Kitton, in Cooke's "Ponds and Ditches," p. 89, etc. (1880.)

FUNGI.

THE old definition of Fungi is now scarcely satisfactory, as it circumscribed them to cellular plants, deriving nourishment by means of a mycelium from the matrix, and never producing from their component threads green bodies resembling chlorophyl. This definition was intended for the distinguishing of fungi from lichens, the latter deriving nourishment from the surrounding medium, and not from the matrix, constantly producing in the thallus green bodies resembling chlorophyl. It is very difficult, in a few words, to give an unimpeachable definition of a fungus. Nor is this very necessary in a book of this kind, since the general notion of a fungus is not so restricted as it was in the past, when only a few prominent types were popularly recognized. It is, however, of some importance that it should be known how widely these organisms are distributed, and what an immense number of species have been recognized by scientific men. It seems but a few years since the botanical world only accepted the possibility of some twenty thousand species being known throughout the world, whilst to-day we possess the knowledge of forty thousand described species, and these are continually on the increase. If we turn back to Lindley's "Vegetable Kingdom," the edition of 1847 only estimates the total number of species of fungi at four thousand, so that in less than half a century the number has been increased ten times, and with that increase has come a deeper and wider knowledge of the mysteries of life as developed in some of the lower forms of the vegetable kingdom.

If we were required to indicate briefly the most prominent types of fungi, we would say that the best known are the large fleshy fungi, with a cap and stem, of the "mushroom and toadstool" form, of which more species are now known than there were of all kinds of fungi known in 1847. Following these are the hard, woody Polypori, which grow on the trunks of decayed trees. All these are large and conspicuous, being included within the popular conception of what constitutes a fungus. Then there is another important group, with a vastly different habit and appearance, known popularly as "rusts" and "smuts" which attack, and are parasitic upon, growing vegetables, causing their destruction. In addition to these are the little cup-shaped objects which occur upon decaying vegetable substances, and sometimes upon the ground, and on dung, a large and widely dispersed group of fungi. A still larger group consists of individuals for the most part like little black dots on the dead stems of herbaceous plants, and on the dead branches of trees, and old trunks. They are numbered by thousands, and the fructification is of a complex character, enclosed within hard, bottle-shaped receptacles, termed perithecia. Beside all these, and some smaller groups not enumerated, there are two rather large sections of imperfect fungi, one of which resembles the last named, except that the fructification is more rudimentary and simple ; and the other consists of the moulds of the type—let us say—of the common blue mould, but which are regarded as the conidia, or imperfect conditions of more highly developed fungi. In order to complete this rapid summary we must add the Myxomycetes, the vegetative system of which seems to have some affinity with the animal kingdom, and the reproductive to be related to fungi. These are still attached to the fungi as an outside group. Finally, there are the very minute, but important, yeast fungi, and the bacteria-like fungi, which seem in one direction to connect the fungi with algæ, many of them intimately related, without doubt, to the epidemic diseases of animals and plants. Such is the very varied assemblage of organisms which constitute the FUNGI.

THE DESTROYER.

The general definition of fungi, which is given in all the text-books, furnishes an excellent clue to the character and functions of these organisms. We allude to the intimation that they "derive nourishment by means of a mycelium from the matrix on which they grow." This is so generally true, and the exceptions more apparent than real, that we may accept the fact that fungi are the destructive agent in organic nature. It is only necessary to indicate a few instances, as types of the manner in which this destruction is accomplished. If we analyze

the large group of gill-bearing fungi—that is to say,
those of the mushroom type—we shall find that nearly
twenty-nine per cent. grow upon decayed wood. In
these cases, life having departed from the wood, the
mycelium penetrates the tissues, disintegrates the
cells, and produces a condition which we call decay,
but which is in effect reducing it to a pabulum
capable of supporting the life of the agaric which is
to be developed from the mycelium. Dead wood,
so long as it remains dry, resists the encroachments
of the fungus; but when persistently moist it soon
falls a prey. All decaying wood is more or less penetrated by fungus mycelium, whether the fungus itself
is developed or not, the full development depending
upon a sufficiency of moisture or other surroundings.
Doubtless continued moisture predisposes the wood
to decay, dissolves what is soluble, softens the cell
walls, and induces a kind of fermentation; the
growing mycelium does the rest by slow disintegration, and the liberation of the chemical constituents,
so that the main factor in the destruction of dead
wood is fungus mycelium. Out of a total of 2711
white-spored agarics no less than 924 are lignicolous.
Of 334 pink-spored species there are 81. In nearly
1000 brown-species, or *dermini*, 218 are found habitually on wood; and of 608 blackish-spored agarics,
the *melanosporœ*, only 98 are lignicolous. Thus, in
4639 species of the Agaricini, 1321 are wood-loving
species, or, in fact, destroyers of wood.

The destructive process is extended also in the
same manner to dead leaves fallen on the ground,
and consequently continually moist; but their final

reduction to vegetable humus is expedited by the growth of fungus mycelium. From the 3318 species left after deducting the foregoing, we must set apart about 235 species which habitually flourish on dead leaves and the stems of herbaceous plants. We thus arrive at an additional seven per cent. of direct fungoid disintegration.

By no means an insignificant number of fungi of the agaric type flourish on dung, almost exclusively such dung as contains a large percentage of vegetable matter, as horse-dung and cow-dung. The large amount of nitrogenous matter in those excrements is eminently suitable to the development and growth of the agarics; but, although mycelium rapidly penetrates the matrix and aids in its disintegration, these species can scarcely be placed in the same rank with the destroyers of wood and vegetable *débris*. Out of the total of 4639 species we recognize some 100 as growing habitually upon dung. There are also a few species which are developed upon decaying fungi which might have been included with the foregoing.

We are now left with a balance of about 2983 terrestrial species, or sixty-four per cent. of the total number of species. These produce a copious mycelium in the soil ; but this mycelium is only in a very limited acceptation destructive. We can scarcely separate those which are known to flourish on old charcoal beds, decaying sawdust, vegetable humus, or even those which ostensibly grow upon the ground, but undoubtedly thrive at the cost of buried vegetable matter, the sites of decayed stumps, or

fragments of old roots. All we can claim for them
is that all these agarics flourish upon their matrix,
deriving their nourishment from the substance upon
which they grow, which must be nitrogenous, and
consist more or less of vegetable or animal matter
diffused through the soil, and not its inorganic
constituents.

Of the residue of the Hymenomycetes little requires
to be said. Nearly all the *Polyporei*, and most of
the *Thelephorei*, grow on rotten wood, which is penetrated by the mycelium. Two familiar species—
Polyporus hybridus and *Merulius lacrymans*, both
known as "dry rot," are in evidence for their power
of destruction.

Pre-eminent above these saprophytes are all those
parasites which attack living plants, and compass
their destruction. There can be no doubt about the
rusts and brands, the whole family of the Uredineæ,
the rust and mildew of wheat, the hollyhock disease,
the plum-leaf rust, all of them determined foes of the
plants upon which they flourish. Who shall estimate
the losses which they are capable of inflicting upon
the cultivator? and yet there are upwards of twelve
hundred different species which attack various plants,
and some unfortunate hosts are the victims of two
or three distinct species, all of which appear to defy
the ingenuity of man to eradicate them.

Equally injurious in their effects and persistent in
their attacks are the "rotting moulds," of which the
potato-disease is one form, the American vine-disease
another, besides many others which are only of less
importance because the plants they attack are less

extensively cultivated, and less associated with the supplies of human food. No one who has had the experience of any of these pests amongst his lettuces, onions, tomatoes, or in his clover field would estimate lightly their powers of destruction.

Another group which may be referred to, in passing, are represented by the hop-mildew and the English vine-mildew. In this family of the *Erysiphei* the mycelium is developed on the surfaces of the leaves, and the host plants are injured by a kind of suffocation, through the obstruction of all the air-passages. The results are no less disastrous and fatal than in those cases where the mycelium penetrates the tissues. About 120 species are known, and all are destructive.

The common fly-mould may be taken as the type of a considerable number of parasitic fungi which establish themselves upon the bodies of living insects, and replace all the tissues by mycelium, which soon causes death. Many of these appear as an epidemic to which hosts of flies and other soft-bodied insects fall a ready prey. Although vastly different in their structure and habits, the aquatic moulds, such as *Saprolegnia*, which is the moving cause of the salmon disease, may be mentioned as carrying on the work of destruction.

Amongst imperfect fungi, such as the moulds (*Hyphomycetcæ*), the majority are probably saprophytic, and only assist materially in the disintegration of dead substances ; but some entire genera are parasitic, and attack living plants. Of these latter we might enumerate *Ramularia* and *Cercospora* as the most extensive and most fatal. It is unnecessary to allude in detail to the yeast fungi (*Saccharomyceteæ*) or the microbes (*Schizomycetæ*), since it is impossible to estimate the destructive influence which they exert over vegetable and animal life, an influence which has only come to be appreciated within the past few years.

The above observations are only offered as suggestive of the almost universal destructiveness of fungi, either upon living tissues or in the disintegration of dead organic matter. On this subject we have written elsewhere[1] to the following effect: "Whenever we encounter decaying vegetable matter we meet with fungi living upon and at the expense of decay, appropriating the changed elements of previous vegetable life to the support of a new generation, and hastening disintegration and assimilation with the soil. No one can have observed the mycelium of fungi at work on old stumps, twigs, and decayed wood without being struck with the rapidity and certainty with which disintegration is being carried on. The gardener casts on one side, in a pile as rubbish, twigs and cuttings from his trees which are useless to him, but which have all derived much from the soil on which they flourished. Shortly fungi make their appearance in species almost innumerable, sending their subtle threads of mycelium deep into the tissues of the woody substance, and the whole mass teems with new life. In this metamorphosis, as the fungi flourish so the twigs decay; for the new life is supported at the expense of the old,

---

[1] M. C. Cooke, " Fungi, their Nature, Influence, and Uses," p. 222. (London, 1875.)

and together the destroyers and their victims return
as useful constituents to the soil from whence they
were derived, and form fresh pabulum for a succeeding season of green leaves and sweet flowers. In
woods and forests we can even more readily appreciate the good offices of fungi in accelerating the
decay of fallen leaves and twigs which surround the
base of the parent trees. In such places Nature is
left absolutely to her own resources, and what man
would accomplish in his carefully attended gardens
and shrubberies must here be done without his aid.
What we call decay is mere change—change of form,
change of relationship, change of composition ; and
all these changes are effected by various combined
agencies—water, light, air, heat, these furnishing new
and suitable conditions for the development of a
new race of vegetables. These, by their vigorous
growth, continue what water and oxygen, stimulated
by light and heat, had begun ; and as they flourish
for a brief season on the fallen glories of the past
summer, make preparation for the coming spring."

On the same subject the Rev. M. J. Berkeley has
written in similar terms.[1] " Fungi," he says, "are
indeed one of the great instruments which keep up
the balance between animal and vegetable life. No
sooner does death take possession in any vegetable
than a host of fungi of various kinds are ready to
work its decomposition. This is at once evident in
all softer structures, which are soon reduced to humus
by the combined action of putrescence and fungi ;

[1] M. J. Berkeley, "Introduction to Cryptogamic Botany," p. 239.
(London, 1857.)

the one, in fact, being frequently the handmaid of the other. The hardest wood, however, yields, though more slowly, to the same agent, and indeed far more rapidly than it would do under the action of mere climatic conditions. A stump of one of our largest trees, if once attacked by fungi, will in a short time present a mere mass of touchwood, which is nothing more than woody tissue traversed and disorganized by mycelium. The same stump, if simply left to the action of the weather, might be half a century before it was fairly decayed. The appearance of such a fungus as *Polyporus squamosus* is the sure harbinger of speedy decay. Nor is the case much mended supposing vegetation still to exist in the stump ; for though the mycelium cannot prey on cells full of vital energy, life is so depressed by the presence and contact of tissues already diseased that the healthiest soon fall a prey to the spreading mycelium. There are, indeed, hundreds of fungi of the most varying size, form, and appearance, which more or less speedily accomplish the same end ; and there is sometimes a host equally fatal to some individual species."

In addition to the foregoing, it is scarcely necessary to do more than mention the name of the " dry rot," which exhibits the destructive ravages of fungi in a somewhat extreme form. Although the mischief is caused in the majority of instances by that which is known as the "dry-rot fungus " (*Merulius lacrymans*), there are other and scarcely less destructive species which work in a similar manner. The ravages they commit in ships, and all kinds of wooden structures, can only be conceived by those who have witnessed them. "I knew a house," writes Mr. Burnett, "into which the rot obtained admittance, and which, during the four years we rented it, had the parlours twice wainscoted and a new flight of stairs, the dry-rot having rendered it unsafe to go from the ground floor to the bedrooms. Every precaution was taken to remove the decaying timbers when the new work was done, yet the dry-rot so rapidly gained strength that the house was ultimately pulled down. Some of my books which suffered least, and which I still retain, bear mournful impressions of its ruthless hand; others were so much affected that the leaves resembled tinder, and when the volumes were opened fell out in dust or fragments." The decay of the wood is partly due to the growth of the fungus, especially the mycelium, which decomposes the tissues, and partly to the moisture which this loosening effects and is the means of introducing into its interior. The germs fall into the chinks of the wood, and germinate, widening the chinks by their growth, and thus permit more moisture to permeate, so that by a continuous repetition of these processes the whole strength of the timber is destroyed. It cannot be too often urged that moisture and exclusion of air are the primary factors in the spread of rot. If by any means all moisture can be got rid of, by establishing a thorough current of dry air, the fungus of the dry-rot is sure to succumb. As soon as a condition of dryness is established the filaments will shrivel and decay. The spores can only germinate in a sufficiently moist atmosphere, the filaments can only grow in the presence of moisture. Cut off the supply of moisture and exhaust all that is present, and fungus growth is conquered. The presence of dry-rot is evidence in itself of the existence of superabundant moisture. No one who has experienced the potency of dry-rot in cellars, of mildew on damp linen, or of mould on jams and fruits, will dispute our claim to place fungi above all other vegetable organisms as entitled to the epithet of " The Destroyer."

LUMINOUS AGARICS.

There are many references in books of travel to luminous agarics, which have been observed in different parts of the world, so that the phenomenon itself can no longer be open to doubt, whatever uncertainty there may be as to its proximate cause. Up to this time the largest number of described species, with a phosphorescent property, have been found in the Australasian colonies. Dr. Bennett alluded more than thirty years ago to their occurrence, or, at the least, to one species which had come under his observation. " It is very common," he says, " in the Australian woods, in the vicinity of Sydney, about the localities of the South Head Road, and among the scrubs and forests on the approach to the headlands of Botany Bay, and emits a light sufficiently powerful to enable the time on a watch to be seen by it. The effect produced by it upon the traveller when, on a dark night, he comes suddenly upon it glowing in the woods is startling, for to a person unacquainted with this phenomenon of the vegetable kingdom the pale, livid, and deadly light emanating from it conveys to him an impression of something supernatural, and often causes no little degree of terror in weak minds, or in those willing to believe in supernatural agencies. I have frequently gathered this fungus, and, on placing it in a dark room, found that it has retained the luminous power for two successive nights; the phosphorescence, becoming fainter in the second, disappears entirely by the third night. This fungus is of a white colour above, and of a delicate yellowish white beneath, varying in size from six to ten inches across its greatest breadth. The whole of the plant shines with a pale, livid, and greenish phosphorescent glow, similar to that which obtains in that very luminous aggregate tunicated mollusc, the *Pyrosoma*." [1] Probably this was *Agaricus nidiformis*, which was met with by Drummond in West Australia; but that is not to be determined in the absence of specimens. Drummond, during one of his botanical trips, was struck by the appearance of a large agaric, measuring sixteen inches in diameter, and weighing about five pounds. This specimen was hung up to dry in the sitting-room, and, on passing through the apartment in the dark, it was observed to give out a most remarkable light. The luminous property continued, though gradually diminishing, for four or five nights, when it ceased, on the plant becoming dry. He says, "We called some of the natives, and showed them the fungus when emitting light, and

[1] George Bennett, M.D., "Gatherings of a Naturalist in Australasia," p. 59. (1860.)

P

the poor creatures cried out 'Chinga,' their name for a spirit, and seemed much afraid of it." Another West Australian species is *Agaricus lampas*, which also grew on the stumps of trees—*Banksia* or *Grevillea*. The stump was at the time surrounded by water. It was on a dark night, when passing, that the curious light was first observed. When the fungus was laid on a newspaper it emitted by night a phosphorescent light, enabling persons to read the words around it; and it continued to do so for several nights, with gradually decreasing intensity as the plant dried up. A smaller species, not more than two inches in diameter, *Agaricus illuminans*, has been found in Victoria, New South Wales, and Queensland; *Agaricus candescens*, in Victoria; *Agaricus phosphoreus*, in Tasmania; and *Agaricus Gardneri*, in Queensland. The last-named species was first discovered by Gardner in Brazil. It was encountered on a dark night in December, while passing through the streets of Villa de Natividate. Some boys were amusing themselves with some luminous object, which at first he supposed to be a kind of large fire-fly, but, on making inquiry, he found it to be a beautiful phosphorescent agaric, which he was told grew abundantly in the neighbourhood, on the decaying fronds of a dwarf palm. The whole plant gives out at night a bright light somewhat similar to that emitted by the larger fire-flies, having a pale greenish hue.

In addition to these, Gaudichaud found *Agaricus noctileucus*, in the Philippines; *Agaricus prometheus* was discovered at Hong Kong, by the United States

Exploring Expedition ; and Rumphius recorded *Agaricus igneus* from Amboyna. In Borneo Dr. Collingwood found a luminous species, which was probably *Agaricus Gardneri*, of which he gives the following particulars : "The night being dark, the fungi could be distinctly seen, though not at a great distance, shining with a soft pale greenish light. Here and there spots of much more intense light were visible, and these proved to be very young and minute specimens. The older specimens may more properly be described as possessing a greenish luminous glow, like the glow of the electric discharge, which, however, was quite sufficient to define its shape, and, when closely examined, the chief details of its form and appearance. The luminosity did not impart itself to the hand, and did not appear to be affected by the separation from the root on which it grew, at least not for some hours. I think it probable that the mycelium of this fungus is also luminous, for, upon turning up the ground in search of small luminous worms, minute spots of light were observed, which could not be referred to any particular object or body when brought to the light and examined, and were possibly due to some minute portions of the mycelium. Mr. Hugh Low has assured me that he saw the jungle all in a blaze of light, by which he could see to read, some years ago, as he was riding across the island by the jungle road, and that this luminosity was produced by an agaric." [1]

There is only one European species which is distinctly luminous, and that is the agaric of the olive,

---

[1] Collingwood, in *Journal of Linnæan Society*, vol. x. p. 469.

Agaricus olearius, which has been fully examined and tested by M. Tulasne, who says, "I have had the opportunity of observing that this agaric is really phosphorescent of itself, and that it is not indebted to any foreign production for the light it emits." Like Delile, he considers that the fungus is only phosphorescent up to the time when it ceases to grow ; thus the light which it projects, one might say, is a manifestation of its vegetation.

"It is an important fact," writes Tulasne, "which I can confirm, and which it is important to insist upon, that the phosphorescence is not exclusively confined to the hymenial surface. Numerous observations made by me prove that the whole of the substance of the fungus participates very frequently, if not always, in the faculty of shining in the dark. Among the first agarics which I examined, I found many the stipe of which shed here and there a light as brilliant as the hymenium, and led me to think that it was due to the spores which had fallen on the surface of the stipe. Therefore, being in the dark, I scraped with my scalpel the luminous parts of the stipe, but it did not sensibly diminish their brightness ; then I split the stipe, bruised it, divided it into small fragments, and I found that the whole of this mass, even in its deepest parts, enjoyed in a similar degree to its superficies, the property of light. I found, besides, a phosphorescence quite as brilliant in all the cap, for, having split it vertically in the form of plates, I found that the trama, when bruised, threw out a light equal to that of their fructiferous surfaces, and there is really only the superior surface of the pileus, or its cuticle, which I have never seen luminous.

"As I have said, the agaric of the olive tree, which is itself very yellow, reflects a strong brilliant light, and remains endowed with this remarkable faculty whilst it grows, or, at least, while it appears to preserve an active life and remains fresh. The phosphorescence is at first, and more ordinarily, recognizable at the surface of the hymenium. I have seen a great number of young fungi which were very phosphorescent in the gills, but not in any other part. In another case, and amongst more aged fungi, the hymenium of which had ceased to give light, the stipe, on the contrary, threw out a brilliant glare. Habitually the phosphorescence is distributed in an unequal manner on the stipe, and the same upon the gills. Although the stipe is luminous at its surface, it is not always necessarily so in its interior substance, if one bruises it; but this substance frequently becomes phosphorescent after contact with the air. Thus I had irregularly split and slit a large stipe in its length, and I found the whole flesh obscure, whilst on the exterior were some luminous places. I roughly joined the lacerated parts, and the following evening, on observing them anew, I found them all flashing a bright light. At another time I had with a scalpel split vertically many fungi in order to hasten their dessication; the evening of the same day the surface of all those cuts was phosphorescent, but in many of these pieces of fungi the luminosity was limited to the cut surface which remained exposed to the air, the flesh beneath was unchanged.

"I have seen a stipe opened and lacerated irregularly, the whole of the flesh of which remained phosphorescent during three consecutive evenings, but the brightness diminished in intensity from the exterior to the interior, so that on the third day it did not issue from the inner part of the stipe. The phosphorescence of the gills is in no way modified at first by immersing the fungus in water; when they have been immersed they are as bright as in the air, but the fungi which I left immersed until the next evening lost all their phosphorescence, and communicated to the water an already sensible yellow tint; alcohol put upon the phosphorescent gills did not at once completely obliterate the light, but visibly enfeebled it. As to the spores, which are white, I have found many times very dense coats of them thrown down on porcelain plates, but I have never seen them phosphorescent.

"As to the observation of Delile that the agaric of the olive does not shine during the day, when placed in total darkness, I think that it could not have been repeated. From what I have said of the phosphorescence of *Agaricus olearius*, one naturally concludes that there does not exist any necessary relation between this phenomenon and the fructification of the fungus. The luminous brightness of the hymenium shows, says Delile, 'the greater activity of the reproductive organs,' but it is not in consequence of its reproductive functions, which may be judged only as an accessory phenomenon, the cause of which is independent of, and more general than these functions, since all the parts of the fungus, its entire substance, throws forth at one time, or at successive times, light." From these experiments Tulasne infers that the same agents, oxygen, water, and warmth, are perfectly necessary to the production of phosphorescence, as much in living organized beings as in those which have ceased to live. In either case, the luminous phenomena accompany a chemical reaction which consists principally in a combination of the organized matter with the oxygen of the air; that is to say, in its combustion, and in the discharge of carbonic acid which thus shows itself.[1]

SEXUALITY IN AGARICS.

One of the most fertile sources of romance and mystification, for a century and a half, has been the reproduction of the Basidiomycetal Fungi, as exhibited in the various attempts to discover sexual organs—a question still unsettled, and, as we conceive, uncertain. How are the spores of agarics rendered fertile, and at what stage does fertilization take place? From analogy it may be argued that some such an act should take place, but hitherto the organs and the process seem to have baffled inquiry. One investigator thought that he had found the representatives of the male organs in the ring and about the stems of the agarics, but in this he was evidently mistaken. Some have sought them on the hymenium, or spore-bearing surface; others on the

[1] Tulasne, "Sur la Phosphorescence des Champignons," *Annales des Sciences Naturelles* (1848), vol. ix. p. 338; M. C. Cooke, "Fungi, their Nature, Uses, etc.," p. 104 (1875).

mycelium, or underground threads. In order to comprehend the different theories, it is necessary to premise that the spore-bearing surface consists of three kinds of elongated cells, which are packed side by side. The most numerous of these are the *basidia* (Fig. 38, *b*), cylindrical cells bearing at the apex four little points, or sterigmata, each of which, normally, supports a single spore. Be-side these are smaller bodies, called *paraphyses* (Fig. 38, *a*), which De Seynes considers to be basidia arrested in their development, or atrophied basidia. Finally, there are clavate or, sometimes, spindle-shaped bodies, larger than the basidia, called *cystidia* (Fig. 38, *c*); and these are regarded by De Seynes as hypertrophied basidia; so that, according to him, all the three are simply forms of the same organ, true or fertile basidia, atrophied basidia, and hypertrophied basidia. Our interest will centre chiefly in the *cystidia*.

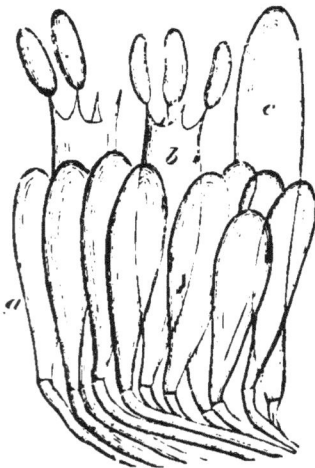

Fig. 38.—*Basidia* and *Cystidia*. *a*, paraphyses; *b*, basidia and spores; *c*, cystidia.

Micheli seems to have been the first who observed particular vesicular organs on the hymenium of *Coprinus*, which were doubtless the cystidia. After-wards Bulliard recognized them, and considered them to be organs of a sexual apparatus, and a sort of spermatic vesicles. Leveille first called them cystidia, but Klotsch and Corda called them boldly "antheridia," anthers, or pollinidia. De Bary describes the cystidia as consisting for the most part of a delicate and colourless membrane, enclosing a similarly colourless plasma, full of vacuoles, and sometimes a perfectly transparent liquid. In *Coprinus micaceus*, he observed a central plastic body, irregularly elongated, which sent in all directions towards the sides of the cell a multitude of filiform processes, branching and anastomosing amongst themselves.

According to Corda, and some other authors, the cystidia eject their contents under the form of a liquid drop, by the summit, which is represented as open. Neither De Bary or Hoffmann were able to convince themselves that this phenomenon was produced spontaneously. If the surface is damp and often bears liquid drops, this is a circumstance which is common to them with all fungoid cells that are full of juice. Those observers who have considered the cystidia to be male sexual organs, have supposed that the ripe and detached spores of the basidia fastened themselves to the moist surface of these cystidia, to be there fecundated by the lubricating fluid. De Bary, it may be observed, did not countenance the idea of the cystidia being anything more than cylindrical hairs, or pilose formations, with no powers of fertilization.

Passing over intermediate observers, as adding nothing definite, we come to the latest exponent of sexuality, Mr. Worthington Smith,[1] who has at least the merit of being more explicit than any of

[1] "Reproduction in Coprinus radiatus," by W. G. Smith, in *Grevillea*, vol. iv. p. 53.

his predecessors.   He at once recognizes the basidium,
or rather the spore, as the female organ, and the
cystidium as the male.   Of the former he says,
"According to my views its analogy is with an unfe-
cundated ovule, without an embryo," and of the latter,
"The cystidium represents with its granules the
anther and its pollen." He describes the cystidia
as sometimes crowned with granules and sometimes
bearing four spicules.   In moisture these spicule-
bearing cystidia germinate at the four points of the
spicules, and produce long threads, which bear at
their tips the granules so frequent in typical cystidia.
The granules at first are not capable of movement,
but are really spermatozoids possessed of fecundative
power.   In certain other of the *Agaricini* the protoplasmic contents of the cystidia are at times discharged from one mouth only, and that at the apex of
the cystidium.   In fluid, after a couple of hours, the
granules, or spermatozoids, begin to revolve, and ultimately swim about with great rapidity.   These spermatozoids attach themselves to the spores, pierce the
coat, and discharge their contents into the substance
of the spore.   From twenty-four to forty-eight hours
after this the spore discharges a cell which soon becomes free, and this is the first cell of the pileus of a
new plant.   The spermatozoids are also represented as
capable of germination, and the production of branching threads, reminding one of a pollen-tube.   We
are not aware of any other reported instance of
spermatia, or spermatozoids, in the cryptogamia,
possessing the power of germination.   The movements of the spermatozoids last for at least four

days. At first they are perfectly spherical, when they merely oscillate ; then they revolve slowly, and, as this goes on, a single turn of a spiral makes itself visible, and the bodies whirl round with great rapidity, but with some intermission. Judging from the presence of the eddy round these bodies whilst whirling, they are probably provided with cilia, but from the extreme minuteness of the bodies themselves their presence cannot be satisfactorily demonstrated. The whirling is sometimes so strong that when they attach themselves to the spores they twist them round.

In the case of *Coprinus radiatus*, and some agarics, the cystidia fall bodily out of the hymenium on to the ground, and it is upon the moist earth that fertilization is generally carried out.

This circumstantial account would furnish a solution of the mystery of reproduction in the agaricini, in great part ; but, unfortunately, although the full details have been before the public for something like seventeen years, we are not aware that they have ever been confirmed by any competent observer. However possible, and even probable, any such revelations may be, they are of such a nature that they cannot be generally accepted as fact without confirmation. Just as well might we accept Ærsted's hypothesis, which is still older, and yet not confirmed. He professed to have seen oöcysts, or elongated reniform cells, springing up like rudimentary branches of the filaments of the mycelium, in a species of agaric, and enclosing an abundant protoplasm, if not a nucleus. At the base of these oöcysts appeared

the presumed antheridia, in the form of slender filaments which turned their extremities towards the
oöcysts, and which, more rarely, were applied to
them. Then, without undergoing any appreciable
modification, the fertile cell, or oöcyst, became enveloped in a lacework of filaments of mycelium, which
form the rudiments of the future cap, the whole
pileus being the result of fecundation. A similar
view was also propounded by Karsten. We are left
with the only alternative of regarding all these as
mere hypotheses until the supposed facts upon which
they are based receive confirmation.

STINKHORNS AND FLIES.

A small but remarkable group of fungi, containing
not more than eighty species, from all parts of the
world, merits notice on account of two or three
characteristics which are probably almost universal.
It might, perhaps, be claimed for them that they are
unique in the number of singular and bizarre forms ;
but form is of less importance than some other prominent features. First and foremost of these is the
possession of a strong and penetrating fetid odour.
It is true that, in some of the species, the descriptions
are deficient in mention of odour, but it is fair to
assume from analogy that a fetid odour predominates.
In a very few isolated instances, amongst thousands
of species of fungi, is fetid odour characteristic ; and,
even in those rare instances, we doubt if the odour
ever approaches in intensity or disgusting fetor those
of the present group. This applies, of course, to

fungi in their healthy and growing condition, and not
to a condition of decay, fermentation, or putrefaction.

The common stinkhorn, or *Phallus*, is often to
be met with in woods and gardens, but no one
attempts to condone or apologize for its odour ; and,
as to its character, there is no variation of
opinion as to its intensity or disagreeable
quality. No one, that we are aware, has
attempted to describe the odour, except
to characterize it as " extremely fetid and
disagreeable." When found growing in a
wood this odour can be detected at some
distance, and long before the object can be
seen, although when diluted by the surrounding air it is not so repulsive as when
in close contact. We remember, some
years ago, whilst on a botanical excursion
with a friend, we discovered a fine *Phallus*,
which had not fully acquired all its intensity of fragrance, and, as it was the first

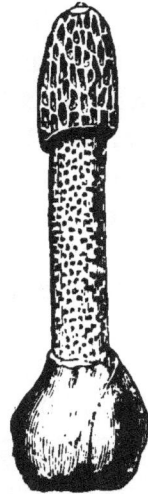

FIG. 39.—Stink-
horn, reduced.

time our friend had seen this object, he was anxious
to convey it home, and make its better acquaintance.
All his sandwiches having been eaten, he resolved to
employ his sandwich-box as a vasculum, chiefly in
order to preserve it from crushing. Thus entombed,
it was conveyed to his pocket, and we thought no
more of it until some time after, when we were safely
ensconced in the train on our way home. Shut up in
a close compartment we soon became conscious that
some of our fellow-travellers were evidently dis-
concerted, and cast inquiring eyes in all directions,
some were even directed under the seats, in order to

discover the source of some odour which was growing
in intensity, and had already become patent to all.
My friend had become painfully conscious of eyes
wandering towards him, but he sat with stolid in-
difference, as if unconscious of any disturbing element,
until, one by one, the passengers vacated our compartment, as soon as circumstances permitted, and
we were left to travel alone. It is not absolutely
certain, but an impression remains that the sandwich-
box was quietly emptied of its contents, out of the
carriage window, before we arrived in town. This
incident will convey some impression of what the
odour of a "stinkhorn" is like, when in perfection.

The species which flourish in tropical countries are
credited with the same abominable smell. The beautiful, but rare *Clathrus*, which has only been found two
or three times in this country, when it escapes from
the volva, exhibits a kind of latticed sphere, bearing
the same dark olive slime of a most fetid character.
Mrs. Griffiths, so well known for her drawings of sea-
weeds, received a specimen of this fungus from Tor-
quay, and proceeded to make a coloured sketch of it
at once ; but it was more than she could accomplish,
for, she writes to Mrs. Hussey, "I was so very much
annoyed with the stench that I could not take more
pains with the drawing ;" the latter lady adds, that
"the same unbearably fetid effluvium escapes from
the deliquescing contents of the network, as in the
stinkhorn." In Australia several species of *Phallus*
have been found, and another rather similar organism
called *Aseröe*, in which the head consists of five or
more rays, expanded like a flower (Fig. 40), of a bright

red colour, the same kind of fetid slimy matter being collected near the centre of the disc. It seems strange, if true, that *Clathrus*, as found in Europe, should be reputed poisonous, and two other species, which at one time were called *Ileodictyon*, should be eaten by the natives in the Australian colonies. Probably,

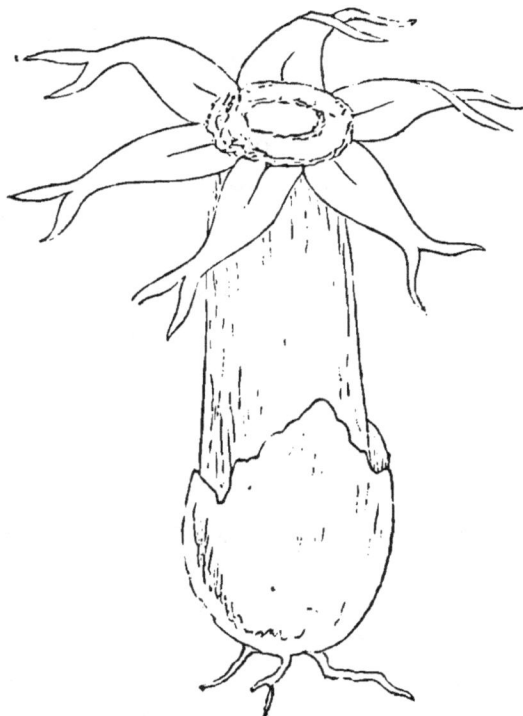

Fig. 40.—*Aserœ rubra*, natural size.

when denuded of the slime, none of the species are injurious, but have acquired that reputation on account of their disgusting odour.

It was in the year 1843 that the announcement was first made that the *Clathrus* had been found growing in Britain. Almost simultaneously it was discovered by Mr. R. Kippist, and by Dr. Bromfield, in the Isle of Wight. The latter found it about the beginning of October, in a damp and grassy hollow, just within and at the bottom of the Pelham Woods by St. Lawrence, where it grew in tolerable plenty over an area of, perhaps, some seventy or eighty square yards or more of sward. "My attention," he says, "was involuntarily drawn to it by the excessively repulsive odour of carrion that pervaded the air around the spot, and which induced one to look about for the *Phallus*, a species well known to emit a scent so analogous to that of a dead animal as to attract, and apparently deceive, the flies, and induce them to deposit their ova on the slimy pileus. The specimens of *Clathrus* I found were many of them as large as an ordinary-sized orange, and presented the appearance of a very coarse or open network, forming slightly collapsed hollow spheres of a bright flesh-red, precisely like raw meat, the vessels of which are still filled with blood; the texture externally cellular, and, I think, laminated, emitting an odour which, unlike that of *Phallus*, was equally offensive, whether near to, or removed from the organ of smell."[1]

In all this group of fungi, the fructifying surface, or portion which bears the spores, becomes a tenacious, slimy, blackish coating, of the consistence of treacle, and in this the greatest intensity of the odour appears to reside. When the common stinkhorn is mature, this slime covers the pileus, and drips from its edge.

[1] "Clathrus in Britain," by W. A. Bromfield, M.D., in *Annals and Mag. Nat. Hist.* (Dec. 1843), p. 451.

In this gluten the minute spores are immersed. As long ago as the seventeenth century it had been observed that this tenacious substance was attractive to flies. In 1826 Dr. Greville said of it : "So very offensive is the smell of this substance that it is seldom allowed to drop away, according to the course of nature, but is generally consumed in a few hours by flesh flies." Thirty years after this Berkeley remarked that "the dripping hymenium affords a welcome food to multitudes of flies." It was, however, not until 1875 that a suggestion was offered as to the interpretation of these insect visitations, in the following words : "This gelatinous substance has nevertheless a peculiar attraction for insects, and it is not altogether romantic to believe that in sucking up the fetid slime they also imbibe the spores, and transfer them from place to place, so that even amongst fungi, insects aid in the dissemination of species." [1]

In order to comprehend the structure and history of the stinkhorn, it will be necessary to give a summary of its principal features and development, almost in the same words as have been adopted by Dr. Wemyss Fulton in reviewing the present aspect of the subject. The hymenophore, or spore-bearing surface, consists, in its earliest stages, of minute swellings, which arise on the underground mycelium. These at first are homogeneous, but gradual differentiation goes on, so that towards maturity the following parts may be recognized : (1) An enclosing

[1] M. C. Cooke, "Fungi, their Nature, Influence, and Uses," p. 123. (1875).

cortical portion, the volva or peridium, composed of three layers, viz., an outer firm skin, or membrane, which is the external peridium; an inner thin membrane which is the internal peridium; and an intermediate, much thicker, layer of translucent, pale yellowish-brown gelatinous material, which is the gelatinous layer. At the base there is a cup-shaped portion which supports the stem, and is continuous laterally by its margin with the layers of the peridium, and below with the mycelium. (2) A central medullary portion, composed of two very different structures; firstly, the gleba, or spore-bearing part, which forms a hollow conical cap, lying within the inner peridium, and surrounding the upper portion of the stem, to the apex of which it is firmly attached. Its outer surface bears the hymenium, and is honeycombed by a large number of irregular chambers, or depressions, in which the mass of spores is lodged. Secondly, the stem, consisting of a cylinder, whose walls at this stage are firm and solid-looking, and composed of a multitude of small, vertically compressed cavities, filled with jelly.

The volva is at first concealed beneath the surface of the soil, but towards maturity it breaks through the ground, and the exposed part gradually becomes conical and finally ruptures, the stem rapidly lengthening and elevating the gleba in the air. These phenomena depend upon a peculiar mechanical change which occurs in the stem. The gelatinous contents of the flattened cavities disappear, and they become excessively dilated, the previously compact stem increasing threefold or fourfold in magnitude, and becoming open and spongy, the distension of the cavities being due to the secretion of air. The protrusion and elevation of the gleba take place with great rapidity, and may be completed in from half an hour to two or three hours, the gleba attaining a height of from five or six to eight or ten inches above the surface of the ground. The utility of this sudden elevation by the substitution of a rapid mechanical process for the slower process of simple growth will be evident in the sequel.

At the time of emergence, and for a brief interval afterwards, the hymenial surface is firm and solid, greenish-grey in colour, and emits a faint, mawkish, but sweetish and honey-like odour, which is attractive to house-flies. Very soon, however, and before the elongation of the stem is completed, it begins to darken, the odour becomes somewhat fetid, and the consistency changes, so that it gets rather sticky and tenacious. A little later it is dark green, almost black, the odour is very strong and repulsively fetid, and its consistency slimy or almost fluid. These changes in the physical character of the hymenial mass begin at the apex of the gleba, and rapidly extend downwards. They seem to depend largely upon the influence of light, for if one side be protected from its action, the change in consistency and colour is retarded on that side. A specimen kept in a darkened place only very partially liquefied, and did not drop off, but dried up into a hard, black, shining, colourless mass. When examined microscopically the fetid fluid is seen to contain myriads of spores, each three micromillimetres long. The rupture of the peridium, and the changes described, occur during the hot months of the year, from the early part of July till the end of September, and therefore at a time when insect life is very abundant, and when myriads of flies abound. As soon as the dung-like odour is developed, the liquefying hymenium is visited by large numbers of flies, which sometimes on hot sunny days almost cover it, and suck up the fluid mass with great avidity, very soon removing the most of it. When the weather is dull, and cloudy or cold, fewer flies are to be seen on the gleba, but it is possible that the deliquescence then goes on more slowly.[1]

The insects which visited the fetid slime were found to be the ordinary blue-bottle, which was abundant, and a large bright metallic green-fly. On empty gleba which had fallen were seen several small beetles and dung-flies. Under further experiment it was found that when the characteristic odour was developed, large numbers of blow-flies rapidly appeared and settled on the deliquescent mass. Flies living for about three weeks on this food died, but no subsequent change occurred in their bodies. By microscopical examination it was demonstrated that thousands of spores clung to the feet and proboscis of the flies. Their excrements consisted at length exclusively of spores, apparently unchanged. It was clear that the flies transported the spores, but whether the latter retained vitality after passing through them had to be determined. To settle this question

[1] Consult "The Dispersion of the Spores of Fungi by Insects," by T. Wemyss Fulton, in *Annals of Botany*, iii., May, 1889.

culture experiments were resorted to, and it was found that, when the flies' excrements were mixed with dung or fecal matter, they germinated and produced a mycelium. Hence the vitality of the spores was not destroyed in their passage through the digestive canal of insects. The inference drawn from the results being, that insects are normally the disseminators of the spores of the stinkhorn, which serves to explain the peculiar liquefaction of the hymenium, and to furnish a purpose for the fetid odour, which is accomplished in the attraction it affords to certain flies.

From the results obtained by investigation of one species of Phalloids, it may be assumed that the deliquescent hymenium of all, and the strong odour of most, have been accounted for, and hence that this small group are pre-eminent above all other fungi in providing special attractions for insects, and in utilizing their services in the dissemination of the spores and the perpetuation of the species. The group, as a whole, belongs to warm regions, and the majority of species inhabit tropical or subtropical countries, where insect life is most abundant. It is also worthy of note that in ninety per cent. of the species the colour of the spore-bearing receptacle is of a most conspicuous colour—either some tint of red, or white, and, when red, of a bright and attractive tint. From tables constructed by Dr. Fulton it appears that seventy-three per cent. of flowers, ninety-six per cent. of Phalloids, and only twenty-four per cent. of other fungi have either white, red, or yellow colouration, which seems to insinuate that bright colouration is of some preponderating service to the Phalloids, and this service may be inferred from the results of the investigations with the common stinkhorn. There is one small fact in connection with this group of fungi which cannot be altogether devoid of importance ; it is the unusually small size of the spores or reproductive bodies, which are uniformly and ridiculously minute, in comparison to the size of the fully developed plants, in all the Phalloids. This minute size must facilitate the admission of the spores into the alimentary canal of such insects as flies, who thrive principally by suction.

Finally, it must be contended that this small group are especially noteworthy, as supreme over all other groups of fungi, in the possession of a strong fetid odour ; in the deliquescence of this fetid portion, which bears some resemblance to putrid animal matter ; in the preponderance of a conspicuous colouration, and in general attractiveness to insects, which consequently swarm about them, and doubtless aid in the perpetuation of the species.

VEGETABLE WASPS.

The romantic story which was told in 1749, and repeated many times afterwards, of the Vegetable Wasps of the West Indies, like all similar stories, had a basis of truth. It was simply an exaggeration and distortion of a natural phenomenon. It is well known to naturalists that in almost all countries there are occasionally found dead insects, in one or other of their stages, upon which a club-shaped fungus is found growing. The plain, unvarnished tale is, that insects of all orders, and in all conditions, whether of larvæ, chrysalis, or perfect insect, are liable to the attacks of a parasitic fungus, which first attacks them most probably in the living condition, and increases and thrives in their interior, at the cost of their tissues, which are ultimately wholly absorbed, and the insect gradually dies, its whole substance, except the outer skin or case, being replaced by the mycelium, or spawn, of the parasite. This condition being arrived at, the external development of the fungus takes place, by the growth of one or more cylindrical or club-shaped processes from some part of the body, usually near the head. These stems vary in size, according to that of the insect, from the thickness of a pin to that of a cedar pencil, and from less than an inch to several inches in length.[1] The apex gradually swells into an oval or cylindrical head, which latter is dotted all over with minute points. When mature, the entire head is covered with little cells, just buried beneath the surface, which contain the fructification of the fungus. This is the full development of the fungus parasite, which appears as an excrescence upon the surface of the insect, and has given occasion for " traveller's tales."

One of the earliest notices of these parasites dates about the middle of the eighteenth century, when some wasps were found in the West Indies with the fungus growing out of them, and under the name of "vegetable wasps" entered the pages of history. This species is a small one, rather tough, pallid, with

---

[1] For full particulars of the various species, consult "Vegetable Wasps and Plant Worms," by M. C. Cooke. (1892.)

a long twisted stem, and a shortly club-shaped head. Perhaps the entire length seldom exceeds one inch, and the thickness not more than that of a good-sized pin (Fig. 41). The discoverer of this phenomenon was a Franciscan friar, called Father Torrubia, who found them at Havana, in 1749. He says, "I found some dead wasps in the fields (however, they were entire, the bodies, wings, and all, and indeed were perfect skeletons). From the belly of every wasp a plant germinated, which grows about five spans high." In 1758 these productions were figured in a book of "Natural History," and represented the wasps as flying away with the fungus attached to their bodies, although the original observer states that he found them dead in the field. Here, then, the distortion of facts may be assumed to have commenced, in the representation of flying insects. Something came also to be added, by way of description, "After they buried themselves in May they began to vegetate toward the end of July, or rather they are found so about that time. When the tree has arrived at its full growth it resembles a coral branch, about three inches high, bearing several little pods, which are supposed by the inhabitants to drop off and become worms, and from thence flies." Thus the original story gained gradually, by repetition, until it became a marvellous tale. The justification of these additions are furnished, it is stated, by Father Torrubia himself, for the original plate to his travels is said by

FIG. 41.—*Cordyceps sphecocephala* on Wasp.

Mr. Gray, who had seen it, to give a representation of
two wasps lying on the ground with a tree growing
out of the base of the abdomen, while three other
wasps are flying round the trees that are growing
from the ground, having a similar tree affixed to
each insect. Each tree is furnished with numerous
trifoliated leaves.

From this region of romance we descend to later
times, to discover what is absolutely authentic in the
story of "vegetable wasps;" and about the year 1824
we obtain a more reasonable and feasible narrative of
specimens, obtained from Guadeloupe, of what is
evidently the same thing, known to the inhabitants
by the name of *la guêpe végétale*, or vegetable wasp.
" The club rises somewhat flexuously or spirally, and
the head, instead of being globose, is ovate." It is
added that the observer noticed "the wasp *still living*,
with its incumbrance attached to it, though apparently
in the last stage of existence, and seeming about to
perish from the influence of the destructive parasite."

A summary of the veritable portion of the above
story has been given in the following words : " Specimens of Hymenopterous insects, resembling wasps,
have been brought from the West Indies, with a
fungus growing from between their anterior coxæ,
and it is positively asserted by travellers that the
insects fly about whilst burdened with the plant.
Upon opening the bodies of the wasps they are found
filled with the mycelium of the fungus, up to the
orbits of the eyes and the points of the tarsi, the
whole of the intestines being obliterated. In such
cases it is to be supposed that the mycelium of the

fungus first kills the wasp by compressing and drying up the body, and then, continuing to grow, occupies the whole of the cavity of the shell of the insect."

There remains only one point upon which it is desirable to comment, and that is the probability or possibility of the insects retaining any life after the external development of the parasite takes place. On this point we hold a decided opinion, based upon an extensive experience. The vegetative portion of the fungus is essentially that which occupies the interior of the insect. From the first period of infection this portion grows and extends itself at the expense of the material upon which it is established. It absorbs and converts the whole of the tissues into mycelium, which gradually replaces the animal structure. So long as there remains any portion of available pabulum for the fungus to assimilate, there is no attempt made at producing fructification. This is a general rule in fungi, of which abundant evidence could be given. So long as vegetation can proceed freely, with the conditions favourable, reproduction is retarded ; but when further vegetation is impossible, through exhaustion of pabulum, attempts are made for the perpetuation of the species. When the insect becomes a prey to the parasite it gradually succumbs, as its tissues become absorbed ; but contemporaneously with this absorption is the obliteration of all animal structure, so that, when complete, nothing but form remains, the whole of the interior having undergone conversion into a vegetable mycelium. Manifestly, under such circumstances no animal life *can* remain. There can be no functions where there are no organs. All accounts and all experiences of the bodies of infected insects agree in the fact that, when carried to the point of external development and the production of the stem, the whole body of the insect is replaced by the fungoid mycelium, and therefore no animal life is possible. Hence it is a myth to describe an insect as moving and flying with a *cordyceps*-fungus adhering to its body, when all organs of locomotion have ceased to exist.

All the romance of "vegetable wasps" being reduced to simple fact, we have nothing left but the parasitism of fungi upon insects, which may assume many forms, and exhibit many variations of structure, but all starting from the infection of living insects with the spores of some previous generation. The forms assumed may be those of the "vegetable wasps," or of the fly-moulds, or of the muscardine of the silk worm, or, lower still, of some microbe ; but the result is the same, in the destruction of the insect. Those persons whose conceptions of fungi are limited to a mushroom, a puff-ball, or a truffle, may possibly be surprised to learn how varied are their forms, and how extended their operations ; but, in none of their phases and in none of their effects are they so marvellous or of such widespread interest, as in their parasitic relationship to animal life. This relationship is becoming more manifest every year, and even now we are probably only at the beginning of far greater and more important discoveries in relation to life, health, and disease.

Before quitting this subject we may allude to the

romantic associations with which the Chinese have
environed a similar insect-parasite to that of the
wasp, but which, in this case, attacks the caterpillar
of some species of moth. This production is called,
in their language, "summer herb, winter worm," intimating that in summer it is a plant, and in winter a
worm or grub. Each individual is about three inches
long, of which about one half is the metamorphosed
body of the caterpillar. It has a great reputation in
medicine in China, more on account of its rarity and
its mystery than of any virtue which it possesses.
Old specimens cost four times their weight in silver.
The narrative of this production is from the pen of
one of the Jesuit fathers, and is to the effect that
"the Chinese suppose that this is a plant during the
summer season, but that in winter its stalk dies, and
the root becomes a worm ; concerning which the
father observed, that nothing could more exactly
express a worm or caterpillar—the head, the eyes,
the feet, and the mouth, being all plainly distinguishable, as well as the several folds and cuttings in of
the body. This account, it is said, was found to be
perfectly true, but the mistake was owing to the
want of proper accuracy in the observation ; for the
body, which was supposed to be the root transformed,
had in reality never been any part of the plant, but
was found to be really and truly a caterpillar.
Some fanciful persons have supposed that when the
time of its change approached it always selected the
roots of this plant as of a proper size and dimensions
for its purpose, and, gnawing off the end, hollowed
away the stump, so as to introduce its tail into the

cavity, where it remained covered with the bark of
the root, which so nicely joins to it that those who
observe it in a slight way cannot but mistake it to
be a part of the root, or the remainder of the root a
continuation of the body. On opening the body of
the larvæ, however, we find that the root of the
fungus entirely occupies the whole interior portion
from the head to the opposite end."

Of these insect parasites, following as types the
West Indian and the Chinese forms, there are many
species known, probably not much less than fifty,
from almost all parts of the world. Some of these
are minute, of which almost the
smallest is found in Europe, on a
dung-fly the stem of which is like
a bristle, and not exceeding half an
inch in length (Fig. 42). In the
Australasian colonies far more im-

FIG. 42.—Dung-fly Club.

posing species are found, growing from the bodies
of dead caterpillars. The clubs are in some cases
simple unbranched clubs, and sometimes these are
divided above, after the manner of a stag's horn.
The caterpillar itself is usually found buried in the
ground, where they retire to undergo their metamorphosis. The fungus appears, more or less, sometimes nearly entirely, above the surface of the soil.
One of the most imposing, and, probably the largest
species known (Fig. 43), was found on the banks of
the Murrambidgee, in New South Wales. The buried
caterpillars reach a length of six inches, and the
parasites are very little shorter when fully matured,
acquiring by age a dark colour. Between these two

extremes occur almost every gradation of size, and

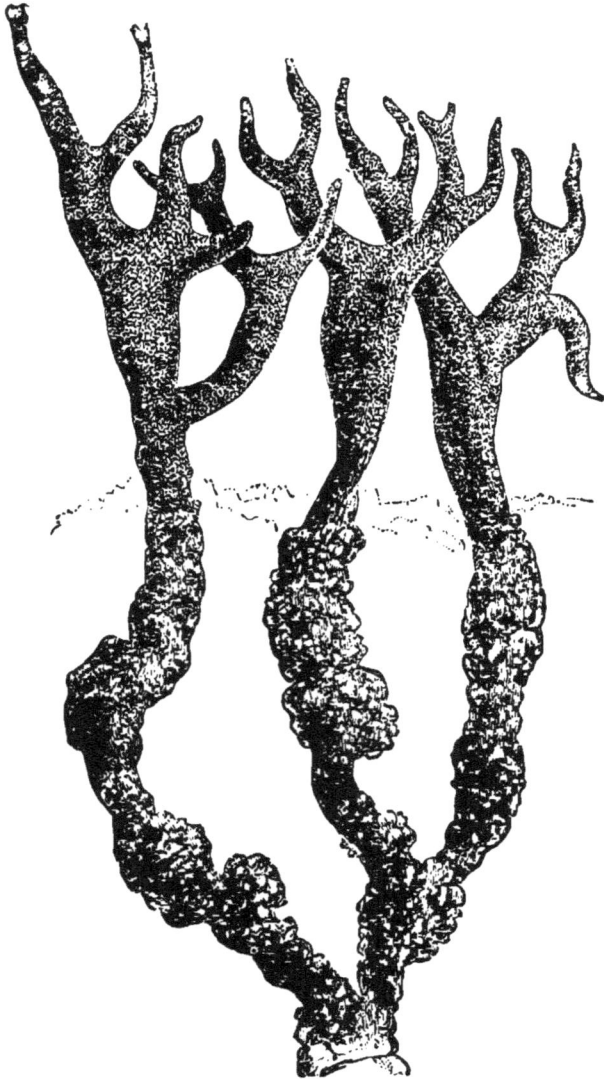

FIG. 43.—*Cordyceps Taylori.*

several modifications of form ; but in all essentials

the structure is the same, the origin the same, and the life-history, with very little variation. There is still one mystery connected with this genus of fungi, on which no light has hitherto been thrown. All the species, in whatever country they may be found, are comparatively rare, and it seems scarcely possible to guess by what means each species is perpetuated. It may be true that each perfect club can develop myriads of spore-bodies, but by what means these are disseminated, and how they find a lodgement in the bodies of healthy insects, is as yet an unsolved problem.

FLY MOULDS.

It is not difficult to impart a superficial idea of what we intend by the term "fly moulds" by calling attention to a very common and well-known example which occurs in the case of the house-fly in the autumn. Especially after moist and oppressive days, flies will often be observed fixed to the window-pane or looking-glasses, as well as other smooth and polished surfaces, by their extended feet, the insects being quite dead, surrounded by a kind of frosty halo resembling a minute snowy-white dust. This frosty appearance extends also to the body of the im-molated fly which has fallen a victim to the fly-mould, *Empusa muscæ*. It is not so very many years since this disease, to which it was then known that flies were subject, was considered to be the only one of its kind, but now some scores are recognized, and all due to the invasion of a parasitic fungus. It was then believed, and even now affirmed by a few people, that when flies killed by this disease fell by accident or were intentionally cast into water, they developed from their bodies another and quite distinct kind of aquatic mould, which was identical with that causing the salmon-disease. If this were a fact the inference would be that, if a fly which had been destroyed by the common fly-mould, *Empusa muscæ*, came in contact with water, the result would be the conversion of the fly-mould into the salmon-moulds, *Saprolegnia ferax* or *Achlya prolifera*. In these latter days such a belief as to the transmutation of species, so distinct and definite in their character, obtains little credit. Whilst the existence of the fly-mould, and also of the salmon-mould, upon flies and other dead insects is not denied.

Our principal concern is with the fly-mould, which may be accepted as the type of certain insect parasites known as the *Entomophthoreæ*, of which a more detailed description is necessary. It may be premised that the frosted condition, and cloudy halo, above alluded to are caused by the external conidia of the mould, by means of which the species is disseminated. There is also, in addition, a mode of internal propagation, by the production of thick-walled resting spores, which, after a period of rest, aid in the perpetuation of the species. So that the fly-mould may be communicated by conidia, and also by resting spores. It has not yet been demonstrated that infection results from the inception of the conidia with their food by healthy insects, but by external application.

The conidium attaches itself to the surface of the

body of a healthy insect, and there germinates. The germinating tube enters the body, either by perforation, or by taking advantage of some natural opening, and when it has entered the body it develops rapidly at the expense of the tissues, producing hyphal bodies, or short thick fragments of variable form (Fig. 44), which continue to increase by budding or division until the body is filled with them, and consequently the death of the insect supervenes. A mass

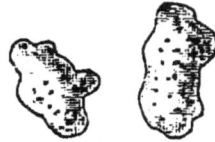

FIG. 44.—Hyphal bodies.

of hyphal bodies being produced within the insect, if the conditions of temperature and moisture are favourable, these commence to germinate. But, if the conditions are not favourable, a resting condition occurs, which may extend for several weeks, until proper conditions present themselves for further development.

Having appropriated the whole of the nourishment which the host affords, and there is a sufficiently moist atmosphere and high temperature, the hyphal bodies germinate with great rapidity; in simple cases producing a thread, which grows directly into the outer air, and develops one or more conidia; in other cases the fertile thread, or conidiophore, is indefinitely branched (Fig. 45). The branching depends very much upon favourable conditions. The threads, which arise directly or indirectly from the hyphal bodies, grow outwards rapidly, burst through the integuments of the dead insect in spongy masses, which are normally whitish, and produce the conidia in profusion. These conidia are terminal on the branches,

R

and are at first produced in a mother-cell, cut off by
a septum from the apex of the conidia-bearer. It is
unnecessary to detail minutely the process of formation and development of the conidia, which would

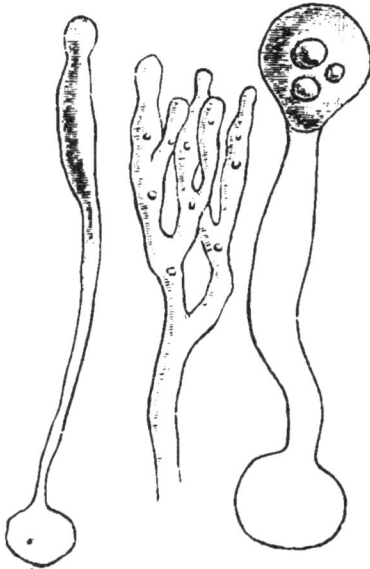

FIG. 45.—Conidia-bearers of Fly-
moulds.

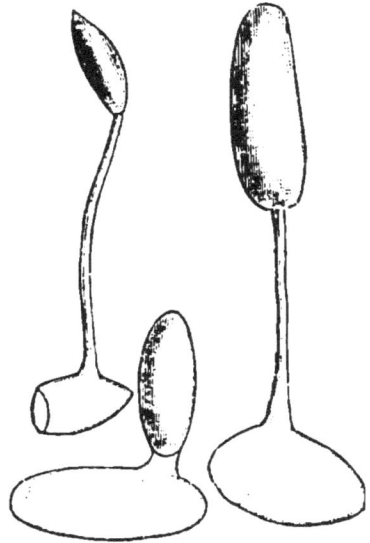

FIG. 46.—Secondary Conidia of Fly-
moulds.

not affect the conclusions to which our remarks are
directed.[1]

When the mature conidium is discharged it may
come in contact at once with a suitable host, germinate, and by means of the germinating thread penetrate and infect a new host ; or, if it falls upon an
unsuitable substance, then the germination results in
the production of a secondary conidium as a provision
for further dissemination (Fig. 46). This secondary

[1] See also "Vegetable Wasps and Plant Worms," by M. C. Cooke,
pp. 8-12. (1892.)

conidium is sometimes able to effect what the primary conidium was unable to accomplish, namely, the infection of some healthy and vigorous insect.

Reverting to the hyphal bodies, which we have followed to the production of conidia, we may observe that *resting* spores are developed also from them within the substance of the dead insect; such resting spores being formed either as the result of a sexual process, and then called *zygospores* (Fig. 47), or without any sexual contact, and then *azygospores*. In the latter case the hyphal bodies

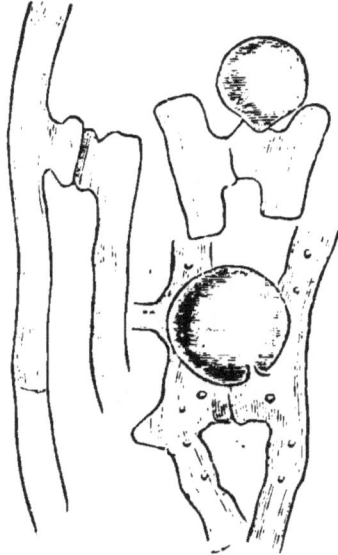

FIG. 47.—Zygospores of Fly-mould.

can be directly converted into non-sexual resting spores by the formation of two additional cell walls, or else by lateral budding from the hyphal bodies. In the case of sexual production, several methods, differing somewhat from each other, have been observed in different species. The most perfect form yet noticed is one in which the hyphal bodies, lying side by side within the host, conjugate in a direct manner (Fig. 48). At first a slight projection appears

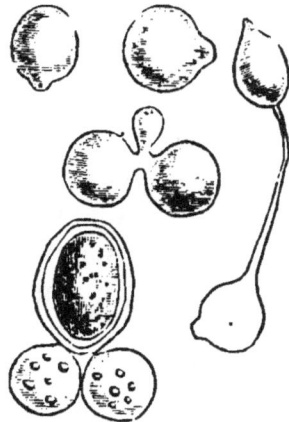

FIG. 48.—Hyphal bodies conjugating.

from the upper end of each hyphal body. These projections soon meet midway between the hyphal bodies, after which a bud begins to arise directly above the point of union. The contents of both bodies pass into this bud, forming the mother-cell of the zygospore. After the spore is mature the remains of the two hyphal bodies are usually persistent for a long time as bladder-like appendages. It has been affirmed that resting spores produced in the autumn germinate in the following spring, but probably the period of rest is very variable. This portion of the life-history requires further elucidation.

These fly-moulds are now known to inhabit various insects, especially two-winged flies ; indeed, all orders of insects have been favoured by their parasitism, but especially we may mention the Aphides, or plant-lice. To what extent artificial inoculation may be possible seems to us at present problematical, since hitherto the infection of fresh hosts artificially has not always been successful and always seems subject to some difficulty. In a state of nature certain of the species appear to be epidemical. Thaxter has observed numerous epidemics of the grasshopper form in the United States.[1] Two he has observed amongst small flies and leaf-hoppers ; and something very like it amongst *Aphides*. In relation to these latter insects it has been observed that the introduction of the parasite may prove a great check, and consequently of enormous benefit in reducing the number of plant-lice so destructive to the hop and greenhouse plants.

[1] "On the Entomophthoreæ of the United States," p. 159.

It is in this direction that a knowledge of the life-history of these fly-moulds may be turned to profitable account, and we do not despair of hearing, at no very remote period, that, by the introduction of infected insects, a kind of epidemic may be produced at will, which will be an important agent in the hands of the cultivator. Mr. Thaxter, writing of *Empusa aphidis,* says that "in greenhouses it acted as a decided check to the multiplication of the *Aphides,* yet did not spread with sufficient rapidity to render smoking in the greenhouse unnecessary." Again he says, "At Kittery I have found it on numerous genera of aphides, and especially destructive to the forms which injure the hop. In one case I observed a large hop vine, some twenty feet high, completely covered with aphides, which were killed off by this fungus in about two weeks ; the affected hosts being fastened to the under sides of the leaves, and to the younger shoots in vast numbers. The destruction of colonies of *Aphis* by this species, and another, seems to be the rule rather than the exception, and is, at least, of very common occurrence. An instance was called to my attention in June, 1886, when I was shown great quantities of aphides dying of the disease on clover near the agricultural department buildings at Washington. The probable agency of ants in spreading these epidemics is worthy of notice, as well as that of night moths, which, as well as ants, are often attracted in great numbers by the sweet secretion of the aphides."

We presume that, having demonstrated that so obscure and apparently simple a form of fungus life as the fly-moulds are in possession of a process of sexual reproduction by conjugation, we have justified a reference to them here. How much more justified do we feel in having made known the various methods by which these simple organisms reproduce and extend themselves amongst a very destructive class of noxious insects ; and, on the romantic side, to suggest, if we can do no more, the inoculation of plant-lice with a potent epidemic, with a view to their destruction, and once more corroborate the axiom that "Knowledge is power."

TRIMORPHIC UREDINES.

Persons who dwell in the country, or are addicted to spending much of their time in rural districts, become acquainted with phenomena in plant life to which citizens are strangers. Certain diseases of the flowering plants, wayside weeds, are familiar to country life by the snuffy-like powder with which the green leaves are sprinkled during summer, which even the uninitiated regard as manifestations of disease, and are apt to distinguish as "rust." As an example of this kind of parasite we will take a common wayside weed called "nipplewort," or botanically, *Lapsana communis.* If the leaves of this plant are examined in March or April, when the plants are still young, some of them will be seen to present irregular purplish spots, perhaps half an inch in diameter on either surface of the leaf, and readily distinguishable by the naked eye. Very soon after their appearance these spots will exhibit, scattered over their surface, whitish dots elevated above the purple substratum, which, by the aid of a

pocket lens, will be seen to have been split at the
apex, with the torn edges turned back like a white
fringe, surrounding a central orifice of a golden-yellow
colour. By using a higher magnifying power, the
yellow centre is found to be caused by a mass of
globose bodies, called spores, or *æcidiospores*, nestling
within the little white fringed "cluster cups," which
latter have their base imbedded in the purple stratum.
The purple spots are appreciably thicker than the
other parts of the leaf, and this thickening is caused
by the mycelium, or rooting filaments of the cluster-
cups beneath the cuticle. Hence, be it observed,
the parasite consists of an innate
mycelium, from which the cluster-
cups are developed, each one consisting of a white cup-shaped *peridium*,
or conceptacle, containing numerous
globose, yellow æcidiospores. If
examined at a sufficiently early
period, these æcidiospores will be seen
to arise from the basal cells of the
cup in parallel chains, the apical one
breaking away consecutively when
mature, and becoming free within the
cup. This is the first stage. The

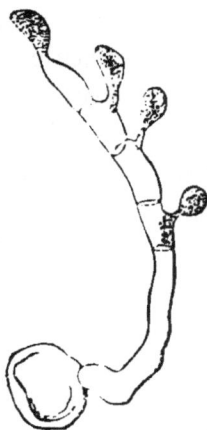

FIG. 49.—Germination
of æcidiospore.

resulting spores are capable of germination and reproduction (Fig. 49).

Later on, from the end of April until early in June,
the same, or other, leaves will exhibit little orbicular
pustules, scattered over the whole surface, each one of
which splits in an irregular manner, and discharges a
cinnamon-coloured powder, which is soon sprinkled

over the leaves, as if dusted with snuff. This powder consists of ovate, or almost globose, pale brown *uredospores*, with a finely warted or delicately spinulose coat. The mass of spores contained in each pustule originates from a sort of cushion, or spore-bed, composed of interwoven threads of mycelium. As they grow and enlarge, the cuticle of the leaf is ruptured, and the uredospores escape. This condition goes by the general name of "rust," and is the *second* stage. When fully matured these spores are capable of immediate germination, each spore having two or more perforations in the outer wall, through which the germinating tube is protruded. The spore-bed of each pustule continues to produce a succession of uredospores for a considerable time. The germinating tube of each spore is capable of penetrating into the host plant through the stomata, and producing there a new mycelium, eventuating in a new spore-bed, and a fresh pustule.

From June to August the same leaves, for the most part, produce smaller and darker pustules, which otherwise resemble those of the uredospores, but the results are different, since the powder, which is dark brown, consists of spores of another type, more elongated, and divided in the centre by a transverse partition into two cells, seated on a colourless stem. This kind of spore is termed a *teleutospore*, and, when each of the three stages was supposed to be a complete and perfect fungus, these bicellular spores were called *Puccinia*, a name now applied to a combination of all three stages. Each cell is perforated by a pore, through which, in germination, the germ-tube is protruded. This third stage, or condition, is popularly called "brand," and completes the cycle of this and similar trimorphic uredines.

The teleutospores are in most cases "resting spores," that is to say, they do not germinate at once, but undergo a period of rest, probably through the winter months, germinating in spring, and then infecting the seedling plants of its proper host, and thus perpetuate the species. When the teleutospores germinate a germ-tube is protruded through the germ-pore, and, as this tube elongates, the contents of the cell in germination pass into the germ-tube, and retract to the upper or apical extremity, leaving the lower portion empty. The free end of the germ-tube is soon traversed by one, two, or three septa, commencing from above downwards. Each cell now formed by this septation sends off a short pointed branch, then the pointed ends dilate and assume an oval or kidney shape, into which the contents of each respective cell pass when the swollen end becomes constricted into a secondary spore, and falls off. Thus two or three secondary spores, or promycelial spores, as they are called, are produced at the ex-

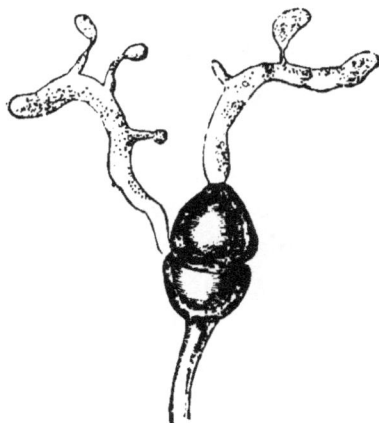

Fig. 50.—Germinating teleutospore of *Puccinia.*

tremity of each germ-tube, from each cell of the teleutospore (Fig. 50). When these promycelial spores

fall upon a damp surface they in turn produce a germ-
tube, supposing that the damp surface upon which
the promycelial spores fall should be a leaf of the
true host plant (in this case of *Lapsana*) the growing
point of the germ-tube enters the cuticle, and commences to form a mycelium in the tissues ; meanwhile
all the contents of the original promycelial spore
passes down into the new mycelium, leaving the
spore-cell empty.

The theory is, that from this mycelium, derived
from the germination of the promycelial spore, a
cluster of *Æcidium*, or cluster-cups, is produced, and
the cycle commences again, æcidiospores, uredospores
teleutospores, *ad infinitum*. From this we learn that
some of the "rust-fungi" pass through three forms or
stages, formerly regarded as three distinct fungi,
namely, *Æcidium*, *Uredo*, and *Puccinia*, a phenomenon
not only curious, but of immense importance in deal-
ing with them as diseases of plants.

It is evident, from the foregoing details, that an
immense power of reproduction is given to these
minute plants. The æcidiospores, by germination,
are able to reproduce themselves directly upon the
host plants, and thus, whatever their influence on
the subsequent uredospores, multiply the sources of
infection. The uredospores, possessing the power of
germination, can transfer the uredo to unaffected
leaves, and thus indefinitely extend the uredo form,
not only upon the same host plant, but to other host
plants of the same species, growing in proximity.
And finally the teleutospores, by remaining at rest
through the winter, provide for the initial infection

of seedlings in the spring by means of the promycelial spores, which by reason of their mode of
development may be three times as numerous as the
original teleutospores. And yet account must be
taken of the latent mycelium which may possibly
remain quiescent through the winter, and resume
activity in the spring, independently of all the other
facilities for the continuance of the species. Surely
in the "struggle for existence" this humble parasite
upon the leaves of an annual herbaceous plant is
amply provided with the means of self-preservation.
We cannot omit to allude to a circumstance narrated
elsewhere,[1] in which it was evident that the seeds
of annuals infected with *Puccinia* could produce
seedlings, all of which were infected by the disease,
without the intervention of promycelial spores, or
any of the direct and ordinary means of infection.

It may be mentioned here, incidentally, that Dr.
Salisbury at one time attributed the disease called
"camp measles" to the corn-mildew (*Puccinia graminis*), and contended that the teleutospores, which
germinated in the damp straw, produced and disseminated the secondary spores in the air, and thus
caused the disease. This is one of the pretty little
speculative romances of the past.[2]

HETERŒCISM.

In order to comprehend the theory of Heterœcism,
it is necessary to bear in mind the basis of fact

[1] W. G. Smith, "Diseases of Field and Garden Crops," p. 182.
(1884.)
[2] M. C. Cooke, "Fungi, their Nature, etc.," p. 213.

supplied in the preceding chapter on the "Trimorphism of Uredines." It will be seen that, in the example given, all the phases were passed upon a single species of host plant, and, whatever form the uredine assumed, that form was developed upon the leaves of the "Nipple-wort." Bearing in mind the fact that the consecutive *Æcidium*, *Uredo*, and *Puccinia* are accepted as three manifestations of the same parasite in different stages or forms, the problem which presented itself for solution was to harmonize with this conclusion the two anomalies, that certain forms of *Æcidium* are constantly being produced upon berberry leaves, coltsfoot, and other plants, with no corresponding *Uredo* or *Puccinia ;* and that also certain forms of *Uredo* and *Puccinia* are habitually developed upon grasses, sedges, and other allied plants, which never produce an *Æcidium*. In order to reconcile these anomalies, the hypothesis of "Heterœcism" has been propounded, which primarily assumes the possibility that the *Æcidium* may be produced on the leaves of one plant, as, for instance, the berberry, and the corresponding *Uredo* and *Puccinia* on another, and entirely different plant, to wit, upon the leaves and culms of cereals and grasses. Now, it is quite clear that an acceptance of trimorphism, as applied to a single-host plant, cannot necessarily involve the acceptance of the theory, that certain of the stages may be passed on one plant, and the remainder upon another, and very different one. Supposing the latter to be true, it must be based upon direct, and not circumstantial, evidence, and the hypothesis must be shown to be sufficient to

account for all the phenomena. As an advocate of
the theory observes, "the fact that a certain number
of Uredines possess the faculty of passing a part of
their lives upon one plant, and the remainder of it
upon another, and a totally different one, is so remarkable that until quite recently there were persons
who declined point-blank to believe it." Premising
that these sceptics *still* exist, we shall content ourselves with a statement of the case, and its evidence,
without detailed arguments in opposition or defence.

The first assumption of heteroecism is that, because
certain of the Uredines manifest themselves under
three forms, *all* should necessarily follow the same
type, especially those which occur on graminaceous
plants. To this it is objected that it has never been
proved that, in any case, the Æcidium form is essential to the production in nature of the Uredo and
Puccinia form. Further, that there are instances of
Uredo and Puccinia forms being commonly produced
on plants, other than the grass family, without the
intervention of any Æcidium form ; indeed, two entire sections, called respectively *Brachy-puccinia* and
Hemi-puccinia, are of this character. Finally, it is
impossible, if such intervention were necessary, for
the limited supply of *Æcidium berberidis* in this
country to influence the enormous production of the
Uredo and Puccinia in corn, or its total absence in
Australia to effect the same result.

The second assumption is that artificial cultivation, under artificial circumstances, affords proof that
the phenomena produced are identical with those
which would occur spontaneously in a state of nature.

For example, it is contended that when the pro-mycelial spores of Puccinia are sown upon berberry leaves, the berberry Æcidium is produced by artificial cultivation. In nature we have seen a berberry bush in full vigour growing in the hedge of a large wheat-field without the Æcidium upon a single leaf. Which fact carries the greatest weight?

If the believers in heterœcism do *not* believe that the berberry Æcidium is an integral part in the life-history of the wheat-mildew, why do they persistently urge the destruction of all berberry bushes in the interest of the farmer? If they *do* believe in it, why have not the enormous preponderance of Puccinia spores long since killed off every berberry bush in Britain by infecting every living leaf?

The method of proof furnished in support of this hypothesis may be gathered from the report of De Bary's experiments in 1865. "He selected the rest-ing spores (teleutospores) of the Puccinia, and having caused them to germinate in a moist atmo-sphere, he placed fragments of the leaves on which they had developed their secondary spores on young but full-grown berberry leaves, under the same atmo-spheric conditions. In from twenty-four to forty-eight hours a quantity of the germinating threads had bored through the walls and penetrated amongst the subjacent cells. This took place both on the upper and under surface of the leaves. Since, in former experiments, it appeared that the spores would penetrate only in those cases where the plant was adapted to develop the parasite, the connection be-tween *Puccinia graminis* and *Æcidium berberidis* seemed more than ever probable. In about ten days the spermogonia appeared. After a time the cut leaves began to decay, so that the fungus never got beyond the spermogonoid stage. Some three-year old seedlings were then taken, and the germinating resting spores applied as before. The plants were kept under a bell-glass from twenty-four to forty-eight hours, and then exposed to the air like other plants. From the sixth to the tenth day yellow spots appeared with single spermogonia ; from the ninth to the twelfth, spermogonia appeared in numbers on either surface ; and in a few days later, on the under surface of the leaves the cylindrical sporangia of the *Æcidium* made their appearance, exactly as in the normally developed parasite, except that they were longer, from being protected from external agents. The younger the leaves the more rapid was the development of the parasite, and sometimes in the younger leaves the luxuriance was far greater than in free nature. Similar plants, to the number of two hundred, were observed in the nursery, and though some of them had *Æcidium* pustules, not one fresh pustule was produced ; while two placed under similar circumstances, but without the application of any resting spores, remained all the summer free from *Æcidium*." [1]

Rotting Moulds.

The most destructive moulds which attack living

---

[1] *Monatsber. Kon. Preuss Acad. Wiss. au Berlin*, January, 1865; *Jour. Roy. Hort. Soc.*, vol. i. n.s. 107 ; "Fungi, their Nature, etc.," p. 199.

plants, and spread dismay amongst cultivators of garden produce, are those which have been called "rotting moulds"—such as the potato-disease, and a similar disease on spinach, lettuces, and green onions. All these several diseases are.eminently destructive, and all are so much alike in their structure and life-history that the description of one may, in all essential particulars, be likewise applied to the others. They are included here in order to demonstrate the elaborate system which is developed in plants so minute, for the continuance of the species, and to give some idea of the difficulties in the way of eradicating or mitigating the disease when it has once obtained a footing. The first external evidence of the presence of these fungi will be found in their sickly appearance, and the speedy appearance of mouldy spots on the leaves and stems, frequently the under surface of the former, resembling patches of meal, but soon revealed to be effused clusters of delicate threads of mould, crowned at their apices with myriads of spores. This is the first condition which presents itself to the naked eye, but before this occurs the great mischief has been done, and the parasite has established itself effectually throughout the tissues of the host plant.

As a starting-point we take one of the spore bodies, or conidia, which are developed in such profusion on every fertile thread of these tufts of mould when mature. This conidium is an elliptical, or subglobose, colourless minute body, having a thin smooth outer coating of membrane, enclosing the fluid contents. These contents, or plasma, soon become granular,

and at length collect at three or four centres, which
condense, and ultimately are distinctly separated
from each other by the growth of a special envelope.
Finally, the membrane of the mother-cell is ruptured,
and the three or four smaller bodies, which have been
differentiated from the cell contents, make their
escape, each one furnished at one extremity with a
pair of delicate movable hairs, by means of which
these little bodies, now termed *zoöspores*, can swim
actively in any thin film of moisture, upon which
they may fall. Possibly this film may be on the
leaf of a foster plant. In a short time all motion
ceases, and the zoöspores come to rest, the pair of
delicate cilia are absorbed, and a germinating thread
is produced, generally from the opposite end of the
now quiescent zoöspore, the point of which seeks out
and enters at one of the stomata, or pores, of the
sustaining plant. It must be observed here that a
power of selection seems to exist, for the point of
the germinating thread will not enter by the stomata
of any plant, indiscriminately, upon which it may fall,
but only upon such leaves as belong to the particular
plant, or species of plant, of which it is the recognized
parasite. Or even, should it enter the pore of an
alien plant, the thread develops no further, and no
infection takes place. Having once obtained an
entrance, the thread grows vigorously, and a little
mass of threads, called a *mycelium*, is soon developed
within the tissues, capable of spreading itself through
the plant which it has infected. In this way young
seedlings are infected by means of conidia from old
and diseased plants, or clean plants of greater

S

maturity become subjected to the disease. In the
next stage we discover that this mycelium has developed erect branched threads, which pass out
through the stomata again into the external air,
sometimes singly, sometimes in tufts. These are
the fertile threads of the mould, which soon produce
a single conidium at the tip of each of the branchlets,
just like the original conidium whence the zoöspores
were developed (Fig. 51). When fully matured each
fertile thread produces a score or more of these conidia,
which fall away when ripe, and then undergo transformation into zoöspores, ready and active, prepared
to pass through the same stages again, and indefinitely multiply the pest. This history represents the
ordinary conidial fructification of the mould, by
means of which it is passed from leaf to leaf and
from plant to plant, until the whole area is affected.
How many of the minute conidia may be transported
to a considerable distance by a breath of wind it is
impossible to say, but it is known that they are
capable of suspension in the air, and that they may
be carried to any spot where there is sufficient
moisture for the conidia to be differentiated into
zoöspores, and afterwards come to rest and germinate. This process takes place in summer and
autumn, but there is yet another means by which
the pest is disseminated in the spring.

The mycelium which flourishes within the substance of the plant infested is capable of producing
larger globose bodies, chiefly within the stems, concealed from external view. These globose bodies
secrete a thick envelope, mostly of a brownish colour,

and after development they remain in a state of rest

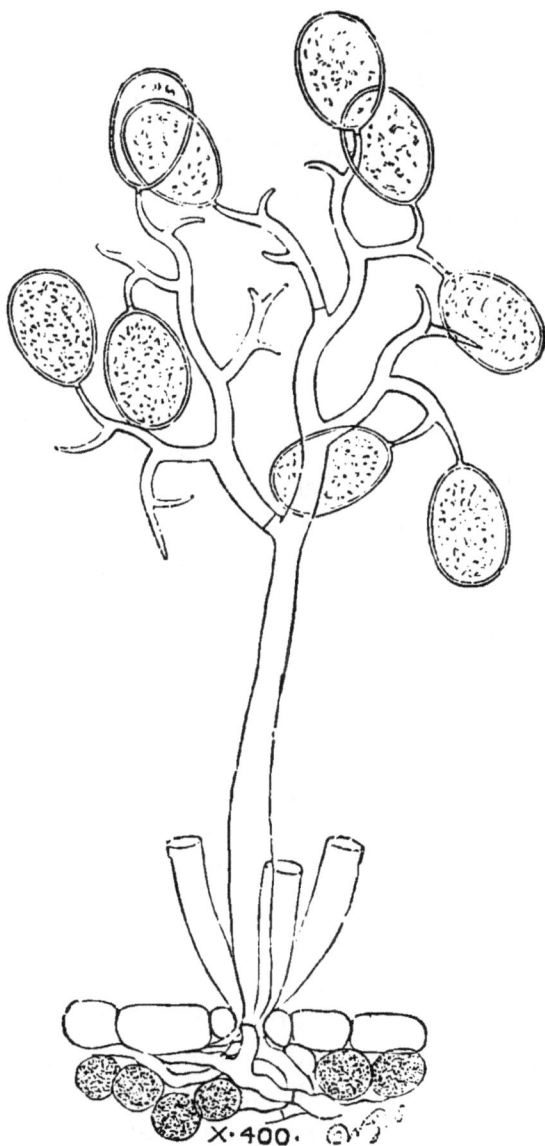

FIG. 51.—Peronospora effusa (*Gardener's Chronicle*).

within the stems during the winter; so that the old stems of plants, which are infested with the mould during the autumn, conceal within themselves, during the winter, a large number of these "resting spores." As the old stems rot and decay, the resting spores are set free in the spring, and then a period of activity commences. The contents of these globose bodies become differentiated into a large number of zoöspores, which ultimately escape by a rupture of the thick envelope, armed with vibratile cilia, and in all respects like the zoöspores which are developed from the conidia. These active zoöspores swarm over the damp soil, and are carried by the spring rains into proximity with the young seedling leaves of the new crop of host plants, then the cilia are absorbed, germination commences, the delicate threads of the mycelium enter the nearest stomata, and infection results. In this way, in addition to the spread of the infection from conidia in summer and autumn, provision is made for an attack upon seedlings in the spring. It will be inferred that, in order to check the spread of these diseases, the conidia must be destroyed in the autumn to prevent their extension to healthy plants, and the destruction of all rotting *débris* must be carried out during the winter so as to extirpate all the concealed resting spores, and thus prevent the infection of seedlings in the spring. Thus it will be noted that knowledge of the life-history of these parasites will suggest the best methods to be employed in their destruction. It is somewhat peculiar that the moulds in this group, when they attack living plants, are rapid in their course of destruction, and that the plants attacked soon rot under the influence of the disease; wherefore the moulds have acquired for themselves the distinction of being termed " rotting moulds."

HIBERNATION.

In the struggle for existence the fungi, of all organisms, seem least fitted to survive the chilling embrace of winter. It is almost matter for surprise that the different species should perpetuate themselves, when it is remembered how, at the approach of frost, so many of them disappear, and seem to leave no trace behind. And yet, year by year, they continue to flourish by some occult power of hibernation ; hence it may prove of some interest to discover how this rejuvenescence is accomplished. In all the gill-bearing fungi of the mushroom type, their cellular structure, and large percentage of water render them very amenable to the effects of cold, so that, after one or two frosty nights, they collapse and disappear ; only a few slimy species, or a few that are tougher and less watery than the rest, survive. When the cap and stem collapse and decay, it would appear to the uninitiated that the whole individual is obliterated ; and, as the spores have a thin membranaceous exterior, that they could not retain vitality throughout the winter. It is still an open question whether and how any of the spores survive. Of the immense number produced by every single individual, possibly very few, under the most favourable circumstances, germinate. Hence, then, as far as we know, or can conjecture, it is not through the medium of the spores, or only to a small extent, that the species is perpetuated through the winter. The only alternative process appears to be by means of a perennial mycelium. It is a well-known fact that the majority of the terrestrial species have the base of the stem attached to a thin delicate network of threads, which interlace the soil, and is called the mycelium, or spawn ; and it is upon this mycelium that the young agarics are developed. The spawn of a mushroom-bed is a case in point. This mycelium we know to be perennial in very many species, possibly in all, and that by resting in the soil, protected and uninjured through the winter, it is capable of reproducing the species in the succeeding year. The occurrence of the same agaric, upon the same spot, year after year for many years, is in favour of this view. In all the "fairy rings" this rejuvenation is doubtless due to a perennial, hibernating mycelium, although it may have originated or subsequently been augmented by germinating spores. The only theory we can propound for the perpetuation of the majority of fleshy fungi is the persistency of the mycelium. Unusually keen winters are generally succeeded by an autumn conspicuous for the small number of individuals of the fleshy fungi, the inference is that the mycelium has succumbed to the frost.

Another view has been maintained amongst a few mycologists, that it is not so much through the mycelium that the species is perpetuated as by the germination of the spores. It is admitted that the spores are difficult to germinate artificially when obtained direct from the mature agaric ; but it is contended that in a natural state they do not germinate directly, but through the intervention of some animal host. The one fact on which this theory is based is that horse-droppings, if collected and placed in favourable conditions, will generate the spawn, and ultimately produce the common mushroom: ergo, the spores of the mushroom will not germinate until they have passed through the intestines of the horse, and that these spores hibernate or germinate and produce a mycelium in horse-droppings during the winter. From this it is inferred that the spores of other species are absorbed by other animals or insects with their food, and thus the perpetuation of the species is secured. One observer, in particular, claims that by feeding some common species of beetle with agaric spores he is able to secure the germination of the spores, and maintain the original species in perpetuity. It would be folly to accept this dictum at once, and before it has been thoroughly tested and confirmed; but it would be equally foolish to deny its possibility, and assert ourselves as judges of possibilities in the obscure phenomena of nature. Hitherto we are wanting in the *facts* upon which such a theory could be established, we recognize many difficulties, and can only hold our judgment in suspense. We are promised to receive some beetles, by cultivation of which we are guaranteed to be able to raise well-known species of agaric, and when this consummation takes place we shall possibly doubt no longer.

Dark-brown leathery threads and plates of a fungoid substance, known collectively as *Rhizomorphs*, have long been known to pass between the bark and trunk of dead trees, and penetrate rotten wood ; but as no fructification could be found, they have been excluded from systematic fungi as a sort of compact mycelium of species unknown. In the light of more recent investigations we feel justified in regarding them as a form of perennial mycelium, or, as we might term it, the hibernating condition of species of arboreal agarics, or of such *Polyporei* as *Polyporus squamosus*, and presumably of *Fomes annosus*. These rhizomorphs are common on timber in mines and places whence light is excluded.

Somewhat allied to *Rhizomorpha* are those hard compact masses of fungus called *Sclerotium*, which present a cellular structure without any trace of fructification. These, again, are of the nature of a condensed mycelium, capable of bearing all the extremes of heat and cold, and possibly in all cases, certainly in a great many, of developing genuine fungi belonging to various genera and species. They are all resting conditions, or hibernating stages, of a very pronounced character. There are already so many of them known that we can only instance a few as typical of the rest. Some of the roots, or rhizomes, of the common wood anemone are liable to become converted into *Sclerotia*, which retain the old form but lose all the old functions. Some persons imagine that they are independent of the anemone, and only grow in its company, whilst we hold it to be a case of conversion, as in the conversion of grain into ergot, and that the *Sclerotium* is a metamorphosed root-stock. After remaining

some time in this quiescent condition, until circumstances are favourable, a cup-shaped fungus (Fig. 52)

FIG. 52.—*Peziza tuberosa.* A, mature cups; B, cellular tissue; C, interwoven hyphæ;
D, asci and sporidia (*Gardener's Chronicle*).

with a long stem (*Sclerotinia tuberosa*) is developed

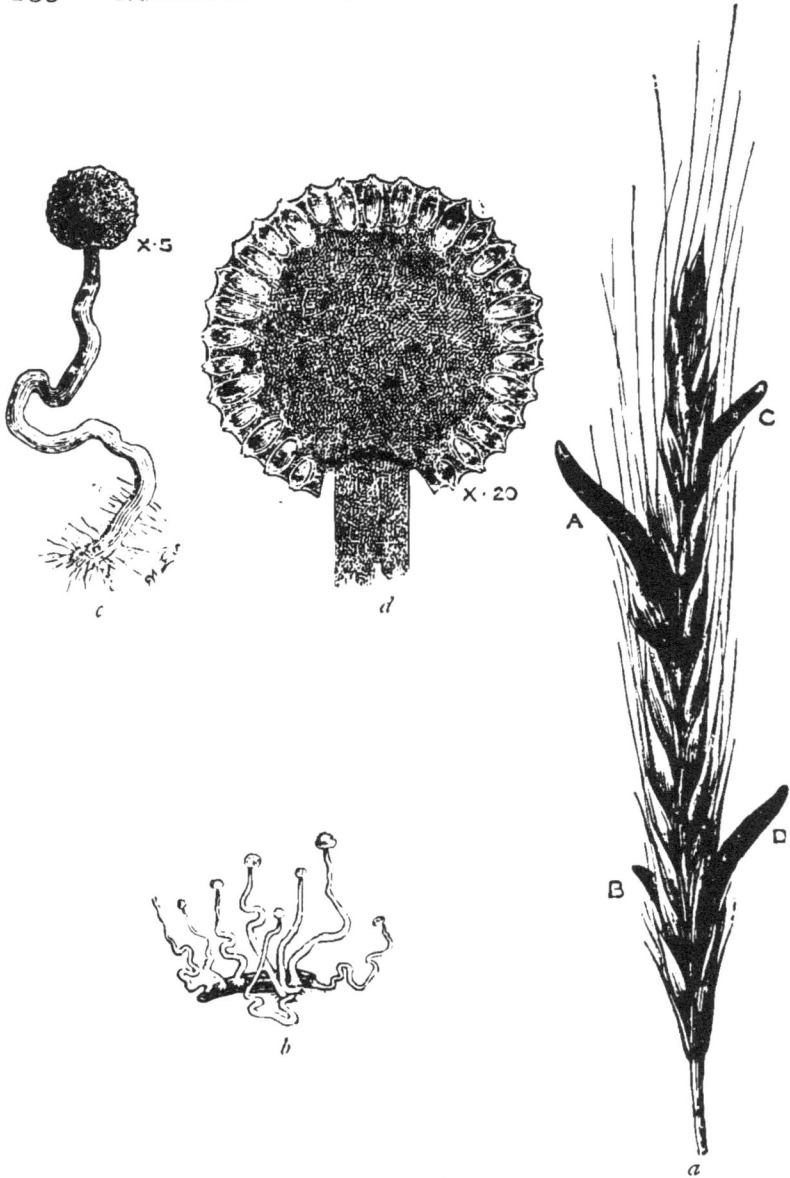

FIG. 53.—*Claviceps purpurea.* *a*, ear of rye with ergol, A, B, C, D ; *b*, *claviceps* grow-ing from ergol, nat. size ; *c*, one *claviceps* enlarged ; *d*, section of fertile head, magni-fied. (After W. G. Smith.)

ledge of this particular fungus, except when springing
from such a compound mycelium.   Other species of
Peziza are developed from other forms of *Sclerotium*.
In the case of ergotized grain (Fig. 53, *a*) the ovary is
converted into a compact *Sclerotium*, which undergoes
a period of rest, perhaps of some months, and then
develops a sphæriaceous fungus called *Claviceps purpurea*.   The ergot is now regarded as the hibernating
mycelium of the more complete fungus *Claviceps*, and
every ergot is held to be capable, under favourable
conditions, of developing one, two, or more of the
little pin-headed fungus.   Cultivators of mushrooms
are painfully aware of the production of a black
Sclerotium in great numbers on their mushroom beds,
to the great detriment of their special cultivation.
These black bodies undergo a resting period, which
may extend through the winter, and are in reality the
hibernating stage of *Xylaria vaporaria*[1] (Fig. 54), which
latter may be cultivated artificially from the *Sclerotium* by placing it in damp sand. The quiescent
condition is developed in autumn, whilst the perfect
state may be looked for in the succeeding March or
April.   The only other example which we purpose
to allude to is one in which the *Sclerotium* is produced in great numbers on old potato haulms, from
the size of a grain of sand to that of a small bean.
After a period of hibernation, and then chiefly by
artificial culture, a small *Peziza* is obtained, growing
from the sclerote.   Although at first considered a new
species, it was afterwards determined to be one which

[1] *Linn. Trans.*, xxiv. t. 25, figs. 17–26.

had appeared on sclerotia very many years previously, and known as *Sclerotinia sclerotiorum* (Fig. 55). It is much smaller than the one above alluded to

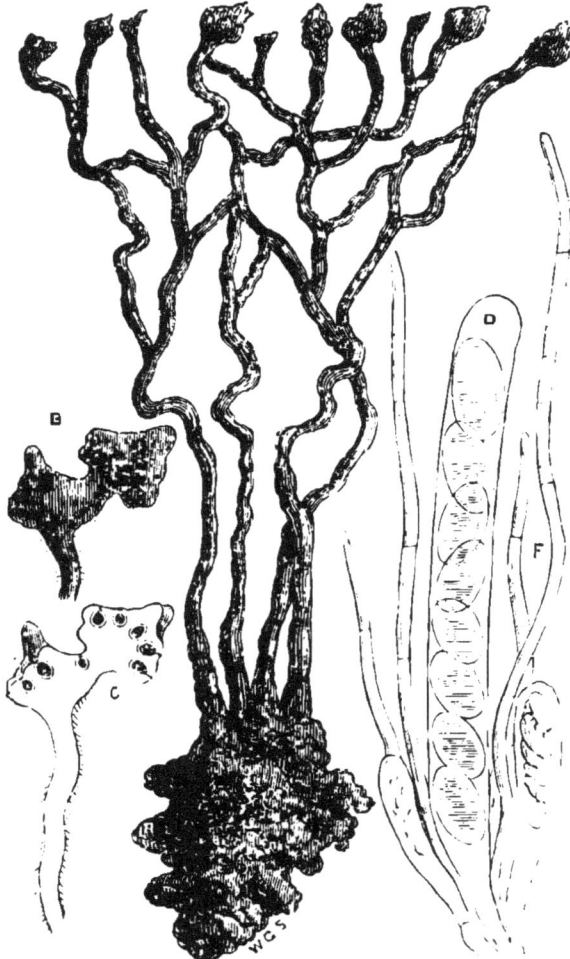

FIG. 54.—*Xylaria vaporaria*, nat. size; B, head enlarged; C, section of same; D, a cis and sporidia; F, paraphyses.

in connection with the wood anemone. Referring to this species, Mr. Worthington Smith makes the following observations.[1] "By means of *Sclerotia* certain fungi, which would probably perish during drought or severe frost, are preserved alive through inclement seasons. The spawn naturally compacts itself into these little hard masses, and falls to the ground; it there remains, like a seed, uninjured by continued

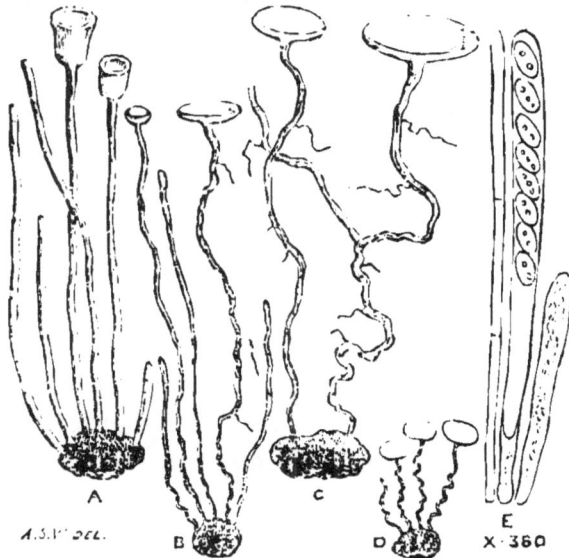

FIG. 55.—*Sclerotinia sclerotiorum*, A, B, C, D, cups, nat. size; E, asci and sporidia magnified (*Gardener's Chronicle*).

dryness or coldness, whereas the vitality of un-compacted spawn would be destroyed under similar conditions in a few hours. Sclerotia vary in their period of hibernation, from a few days to a year or more: they are somewhat erratic as to the required amount of rest; favourable circumstances will hasten germination, whilst unfavourable ones will retard

[1] W. G. Smith, "Diseases of Field and Garden Crops," p. 19. (1884).

it; a very common period of hibernation is from nine months to a year. When germination at length takes place a perfect fungus is produced; this perfect fungus at maturity produces spores, which on germination again produce spawn or mycelium. Sometimes this spawn will at once reproduce the perfect fungus, but in other instances it grows profusely, and at length gives direct rise to the little resting nodosities just described as *Sclerotia.*"

Hibernation is also common with the Uredines. The teleutospores in many species, such as the wheat-mildew, for instance, lie dormant through the winter, but in the spring germination commences, and secondary, or promycelial, spores are produced, through which contact is established with young host plants, a new cycle commences, and the species is perpetuated. There are some species of *Puccinia*, etc., in which the teleutospores are capable of immediate germination; but there are others in which the teleutospores are true resting spores, and lie dormant through the winter.

Some fungi, which manifest themselves externally as moulds, parasitic upon living plants, have a provision, in the production of resting spores, for passing through the winter, so as to be ready for the infection of the young host plants which are developed in the spring. This group of parasites, such as the potato-disease, the onion-mould, and others of the *Peronosporaceæ*, produce thick-coated resting spores upon the mycelium in the interior of the host plant. During winter, as the herbaceous plants of the preceding summer decay and rot, they contain within them these resting spores, which are set free and recover activity in the spring. At this period the resting spores are broken up into a number of small active zoöspores, which float in a film of moisture until a favourable opportunity, when they settle down to germinate and commence a new cycle.

In a similar manner the true Mucors and the *Saprolegniaceæ* pass through a period of hibernation, in which thick-coated zygospores perform the function of resting spores.

The very large section of fungi known as the *Pyrenomycetes* have their reproductive organs enclosed within hard or tough capsules, or perithecia, which are produced for the most part in the autumn upon dead leaves, stems, or branches; and during the winter these remain on the damp ground, maturing their fruit, which is in full perfection in the early spring. In these cases the entire fungus passes through the winter, not in a condition of absolute rest, but of reproductive development. The maple leaves, disfigured with large pitchy black patches in the autumn, must lie on the damp ground through the winter in order to perfect their ascigerous fruit, and then known as *Rhytisma acerinum;* but the same patches in the autumn contain within them certain minute stylospores, not enclosed in asci, and in this condition it has been described as *Melasmia acerina.* The fungus with stylosporous fruit is essentially the same fungus as that which later on has sporidia enclosed in asci. The autumnal stage, or *Melasmia,* was an imperfect condition; the spring stage, or *Rhytisma,* is the perfect form, which has been developed from the former during a period of apparent rest. In this manner thousands of species, in which the reproductive elements are protected from the inclement season by enclosure within sufficiently thick and rigid capsules, lie in apparent rest, while growth and development is slowly and gradually going on.

Of such incomplete fungi as the *Sphæropsideæ* and the greater portion of the *Hyphomycetes* no generalization is at present possible. Possibly the majority of species are only conditions, or stages, of more perfect forms; but their life-history is too little known for speculation. It is assumed that many of the moulds are conidial forms of species of *Pyrenomycetes*, but the true nature of the connection is a mystery. It is a problem how many of the delicate moulds which appear year after year can have passed through the winter, and how the continuity of the species is assured. From what is already known in other sections we have faith in the future, and that what is dark and mysterious to us now will one day become clear and manifest. Of one thing we may be absolutely certain, that it is not haphazard, or chance, in which the destinies of these minute organisms are involved, but that their preservation and development are as well provided for as those of whose life-history we have obtained a knowledge.

NATIVE BREAD.

Several substances have been known, in different parts of the world, under the name of Native Bread; but the organism which is known in Australia and

Tasmania under that designation is the *Mylitta australis*, a kind of *Tuber*, or rather, resembling a *Tuber*, attaining to the size of a child's head, and with a taste somewhat like boiled rice. Mr. Backhouse says, "I have often asked the aborigines how they found the Native Bread, and have universally received the answer 'a rotten tree.'" It resembles a truffle in its more or less globose or irregularly globose form, in its dark colour externally, and in its growing beneath the surface of the ground. Some persons have suspected, or asserted, that it is produced from the roots of some tree. When dried it becomes almost as hard as a stone, so as not to be cut with a knife; but it may be sawn across, and is found to be yellowish white within, and somewhat mottled. No investigation with the microscope has been successful in the discovery of any kind of fruit, so that really it resembles, more than anything else, those compact masses of fungus mycelium, known by the general name of *Sclerotium*. Many of these species of Sclerotia have been induced, under favourable conditions, to develop perfect fungi of various genera from their substance, and recently it has been discovered that a new and true species of Polyporus, named *Polyporus Mylittæ*, is developed, as a final stage, from the Mylitta, or "native bread." In Queensland *Lentinus cyathus* has been found growing upon a sclerotium similar in character and almost as large. A common Sclerotium in mushroom beds produces a club-shaped *Xylaria*, and some small species are the basis of forms of *Peziza*.[1]

[1] See also *Linnean Transactions*, vol. xxiii. p. 93.

T

Berkeley, in 1848, expressed the opinion that the *Mylitta* was closely allied to the common truffle, but we fail to recognize the evidence on which he so regarded it, and now know that he was in error. Tulasne, on the contrary, appears to regard its affinities with sclerotium, whilst neither Tulasne, Corda, Berkeley, nor any of the most distinguished of mycologists who have closely examined this substance, have found structure at variance with or in advance of *Sclerotium*. In support of this view the discovery of the above-named species of Polyporus upon the *Mylitta* may be cited, as certain *Pezizæ* are developed on small species of Sclerotium.

TOUCHWOOD.

In our schoolboy days we have still a vivid recollection of first becoming acquainted with "touchwood," perhaps only a local name, but one which was applied to soft rotten wood which possessed luminosity. Boys are apt to pick up and remember traditions of such phenomena, and we had often heard of marvellous properties of "touchwood," and therefore recognized it at once from the description. It was the hollow of an old tree into which we had penetrated, as a hiding-place from our fellows, some dark evening, and were surprised upon discovering that the chips when disturbed exhibited a distinct pale phosphorescent light. Naturally we took such pieces as exhibited the luminosity most strongly, and pocketed them for further experiment. When in bed the "touchwood" was taken from our pockets and tested, under the bedclothes, after the lights had been taken away, and we, two boys, were left alone. Truly the chips were phosphorescent, with a light strong enough to outline their form, and distinguish the letters upon any book or printed paper on which the fragments were laid. This luminosity was retained for a few nights, and then gradually disappeared. What is now remembered of the wood itself is that it was very soft, crumbling easily between the fingers, thoroughly decayed, and probably deeply penetrated with delicate threads of fungus mycelium.

This circumstance was recalled to memory by reading, many years after, an account given by the Rev. M. J. Berkeley of a similar experience. It was to the following effect: " A quantity of wood had been purchased in a neighbouring parish, which was dragged up a very steep hill to its destination. Amongst them was a log of larch or spruce, it is not quite certain which, twenty-four feet long, and a foot in diameter. Some young friends happened to pass up the hill at night, and were surprised to find the road scattered with luminous patches, which, when more closely examined, proved to be portions of bark, or little fragments of wood. Following the track they came to a blaze of white light, which was perfectly surprising. On examination it appeared that the whole of the inside of the bark of the log was covered with a white byssoid mycelium of a peculiarly strong smell, but unfortunately in such a state that the perfect form could not be ascertained. This was luminous, but the light was by no means so bright as in those parts of the wood where the spawn had penetrated more deeply, and where it was so intense that the roughest treatment scarcely seemed to check it. If any attempt was made to rub off the luminous matter it only shone the more brightly, and when wrapped up in five folds of paper the light penetrated through all the folds on either side, as brightly as if the specimen was exposed ; when again the specimens were placed in the pocket, the pocket when opened was a mass of light. The luminosity had now been going on for three days. Unfortunately we did not see it ourselves until the third day, when it had, possibly from a change in the state of electricity, been somewhat impaired ; but it was still most interesting, and we have merely recorded what we observed ourselves. It was almost possible to read the time on the face of a watch, even in its less luminous condition. We do not for a moment suppose that the mycelium is essentially luminous, but are rather inclined to believe that a peculiar concurrence of climatic conditions is necessary for the production of the phenomenon, which is certainly one of great rarity. Observers, as we have been of fungi in their native haunts for fifty years, it has never fallen to our lot to witness a similar case before, though Professor Churchill Babington once sent us specimens of luminous wood, which had, however, lost their luminosity before they arrived. It should be observed that the parts of the wood which were most luminous were not only deeply penetrated by the more delicate parts of the mycelium, but were those which were most decomposed. It is probable, therefore, that this fact is an element in the case, as well as the presence of fungoid matter."[1]

We have certainly found amongst a rural population many persons to affirm that they have met with luminous decayed wood, so that with them "touch-wood" is not a myth, but a reality, the two essential features of which are—softness, amounting to fria-bility, and the presence of phosphorescence. It must not be assumed, however, that the latter is to be accepted literally, and strictly, as implying the presence of phosphorus in any form, but emitting a pale lambent light, like that which is produced by phosphorus when rubbed.

A curious coincidence is narrated by Sir J. D. Hooker, as having occurred to him during his Hima-layan travels, which must be quoted. "The pheno-menon of phosphorescence," he writes, "is very conspicuous on stacks of firewood. At Dorjiling, during the damp warm summer months, at elevations of five thousand to eight thousand feet, it may be witnessed every night by walking a few yards in the forest—at least it was so in 1848 and 1849, and, during my stay there, billets of decayed wood were repeatedly sent me by residents, with inquiries as to the cause of their luminosity. It is no exaggeration to say that one does not need to remove from the fireside to see this phenomenon, for if there is a partially decayed log amongst the firewood, it is almost sure to glow with a pale phosphoric light. A stack of firewood, collected near my host's cottage, presented a beautiful spectacle for two months (in

[1] *Gardener's Chronicle*, 1872, p. 1258.

July and August), and on passing it at night my pony was always alarmed by it. The phenomenon invariably accompanies decay, and is common on oak, laurel, birch, and probably other timbers; it equally appears on cut wood and on stumps, but is most frequent on branches lying close to the ground in the wet forests. I have reason to believe that it spreads with great rapidity from old surfaces to fresh-cut ones. That it is a vital phenomenon, and due to the mycelium of a fungus, I do not in the least doubt; for I have observed it occasionally circum-scribed by those black lines which are often seen to bound mycelia on dead wood, and to precede a more rapid decay. I have often tried, but always in vain, to coax it into developing some fungus, by placing it in damp rooms, etc. When camp-ing in the mountains I have caused the natives to bring phosphorescent wood into my tent, for the pleasure of watching its soft undulating light, which appears to pale and glow with every motion of the atmosphere; but, except in this difference of intensity, it presents no change in appearance night after night. Alcohol, heat, and dryness soon dissipate it; electricity I never tried. It has no odour, and my dog, who had a fine sense of smell, paid no heed when it was laid under his nose."[1]

SLIME FUNGI.

The peculiar organisms which have been called "slime fungi" have another and more generally

[1] J. D. Hooker, M.D., "Himalayan Journals, etc.," vol. ii. p. 158. (1855).

known name of Myxogastres, or Myxomycetes, which expresses the same meaning. It must be admitted that there are some who deny that they are fungi at all ; whilst others contend that they are an outside group with a little relationship to fungi. For our present purpose we will assume that they are a small group of fungi with an abnormal history. Like many fungi, the spores are enclosed in an integument, more or less dense, until the period of maturity arrives, when the integument is ruptured and the spores escape. Their most characteristic feature is the circumstance that in their earliest stages, or vegetative condition, they consist of a homogeneous gelatinous plasma, which is peculiar to this group alone.

The spore, or individual unit, from which the plants are developed, is minute, and commonly of a globose form. When matured, and circumstances are favourable to their development, either one of the following modes of evolution takes place. The spore originates from its interior one or more movable zoöspores, with vibratile cilia, which ultimately subside into an extensible plasma ; or, from the spores directly, and by confluence, a gelatinous plasma is formed, which bears some resemblance to sarcode, cells of which possess the power of constantly changing their form and moving about in a manner which reminds one of *Amœba.* It is this peculiarity which belongs to the vegetative stage, having no parallel in any other of the groups of fungi, that led superficial observers, fond of theory, to start the hypothesis that the whole of these organisms belong to the animal kingdom, and applied to them the name of *Mycetozoa.*

For some time the hypothesis appeared to maintain its ground ; it was accepted, and advocated in all directions, by men in all departments of biological science, except, perhaps, those who would be expected to be best acquainted with the subject—the mycologists— who did not accept the dictum of the theorists, and did not expunge the *Myxomycetes* from the place they had so long occupied, in close relationship with fungi, although some zoologists proceeded to annex them to the animal kingdom. Undoubtedly, as will be hereafter seen, whilst the vegetative stage has no analogue in the vegetable world, so the reproductive stage has no parallel in the animal world. To quote the most recent authority, "In the Myxogastres the life of an individual consists, under normal conditions, of two very sharply defined stages ; first, the vegetative phase, concerned with functions tending towards the wellbeing of the individual ; second, the reproductive phase, concerned entirely with the continuation of the species."[1] It is shown also, in the same work, that the whole theory of animal nature is based upon the phenomena of the vegetative phase, which are thus summarized : "The spores on germination give origin to one, two, or more naked cells, which possess the power of movement due to the protrusion of pseudopodia, or the presence of a cilium ; these cells are known as *swarm-cells*. The swarm-cells possess a nucleus, multiply by bipartition, and eventually coalesce to form a *plasmodium*, in the following manner. After the production of numerous

[1] George Massee, "A Monograph of the Myxogastres," p. 3. (London, 1892.)

swarm-spores by repeated bipartition, little groups are formed by the close approach of two or more of these bodies ; these groups often disperse again, but eventually the components of the group coalesce and lose their individuality ; this coalescence and loss of individuality results in the formation of a small plasmodium, which, in some unknown way, possesses the power of attracting surrounding free swarm-cells ; these at once coalesce, and add to the bulk of the plasmodium. The nuclei of the component swarm-cells retain their individuality in the plasmodium, the latter retaining the power of motion originally possessed by its components, and represents the vegetative phase of a Myxogaster."

The nearest analogy to phenomena of this kind in cryptogams is presented by the life-history of that pretty fresh-water alga, *Volvox globator*. Dr. Braxton Hicks observed this peculiarity in the cell at an early period, before the young volvox was fully grown, at the time when the future zoöspores first appear, enclosed in cells, the final product of segmentation. These zoöspore-containing cells, by contact with their neighbours, are rendered many-angled, and include from twenty to thirty hexagonal young zoöspores, in close contact, and which are of many colours. When these cells are detached they become round, and have a curious power of changing shape, like a Proteus, protruding the wall first on one side and then on the other, into which protrusions the contents run. The other and more striking instance, however, was visible in the zoöspores themselves at an advanced age, when some of them enlarge and become irregular in outline, some disappear, some break up and disperse within the volvox; some undergo a process of subdivision, producing a group of from two to forty green drops, arranged so that their apices, with cilia, point externally; while others enlarge to two or three times their natural size, having many nuclei within, and variously coloured. When this cell, probably by the solution of the outer mucilaginous coat, becomes free, it also possesses the power of moving precisely as does a true *Amœba*. And yet the suggestion that they are animal cannot be entertained; but the explanation is offered, that the protoplasmic contents, when deprived of their confining envelope of cellulose, possess, in common with sarcode, under certain circumstances, a power of spontaneous motion in the manner of an *Amœba*.

In preparation for the final stage, the reproductive phase of the Myxogaster, each individual acquires a peridium, or outer envelope, and the contents lose gradually their mucilaginous character, and become differentiated into threads, which constitute the capillitium, and a pulverulent mass consisting of globose spores. Nothing in this stage calls for particular comment; it is almost analogous, on a small scale, to the peridium, capillitium, and spores of a *Lycoperdon*, or some other of the *Gastromycetes*, so that no one has ever attempted to deny or explain away its fungoid character. In its reproductive phase the *Myxogaster* is a fungus of the *Gastromycetes*, or puffball type.

It is important in this connection to note an observation made by Meneghini, when advocating the animal nature of Diatoms. "If there are beings," he says, "which, when they attain their perfect develop-ment, prove themselves to be decidedly vegetable, although during the first portion of their existence they presented some phenomena of animal nature, this proves that those phenomena do not exclusively belong to animals, and that we cannot draw from them absolute characters of animal nature. The im-perfection which we have already shown to be inherent in every notion we can form of the animal or vegetable kingdom, begins to diminish after such considerations, and it is under this point of view that we undertake the examination, carefully separating, in the characters they present, those which they hold in common with vegetables from those in common with animals, and inquiring if they do not possess some exclusively with one or the other which may decide the question."

Undoubtedly, in so far as the Myxogastres are concerned, it is admitted on all hands, that in their perfect development they prove themselves to be decidedly vegetable, although during the first portion of their existence they presented some phenomena of animal nature. The inference therefore is, that the phenomena of the first stage do not belong exclusively to animals, and by no means decides the animal nature of the Myxogastres, which should be decided on the faith of their ultimate development.

CELL MULTIPLICATION.

Figures are useless to convey any impression of the great rapidity and endless multiplication of cells in fungi. Mr. Worthington Smith made some interesting calculations on a minute species of *Coprinus*, which he subjected to close examination, but the figures he quotes are beyond human realization. He says, "It will be found that instead of thousands it really requires millions of individual cells to build up one of these minute plants which a breath destroys. The smallness and lightness of one fungus is such that it requires 150 specimens to weigh a grain, 72,000 to weigh an ounce troy. In specimens of *Coprinus radiatus* there were 22,500,000 cells in its structure, irrespective of the spores, which numbered about 3,200,000 more. If all these cells and spores are only equivalent to the hundred and fiftieth part of a grain, it follows that in an ounce of fungus cells there must be no less than one billion six hundred and twenty-four thousand millions of these bodies, exclusive of the spores. In a large mushroom the cells would number hundreds of billions. Still more wonderful is the fact that each individual cell is furnished with a spark of life, contains water, protoplasm, and other material, and is capable of growth and assimilation."

Again he writes, concerning the rapidity of production, that, "if the 22,500,000 cells which go to make up one of these minute plants require fourteen days for their production, it follows as a necessity that the cells go on multiplying all the fortnight, night and day, at the rate of 1·114 to the minute. It takes about five hours for the spores to be gradually produced all over the hymenium—say from five to ten o'clock in the morning,—and as there are upwards of three millions of spores to each plant, they, as a consequence, gradually appear upon the basidia, or spore-bearing spicules, at the rate of one hundred thousand every minute."

Although the calculation was made many years ago, and by what process we do not know, yet the venerable Fries calculated the number of spores, which are simple cells, that are developed in a fungus which seldom exceeds two inches in diameter. He says, "The sporules are so infinite,—in a single individual of *Reticularia maxima*, I have reckoned above ten millions—so subtile they are scarcely visible to the naked eye, and often resemble thin smoke, so light, raised perhaps by evaporation into the atmosphere, and are dispersed in so many ways, that it is difficult to conceive a place from which they can be excluded." We have also in remembrance some calculations made on the production of spores by the giant puff-ball, which are far more astounding, but the reference has hitherto evaded our search. Suffice it to say that the formation of cells in the soft fleshy fungi, under favourable conditions, are so enormous as almost to stagger our faith in figures.

> " Thou art in small things great, not small in any :
> Thy even praise can neither rise nor fall.
> Thou art in all things one, in each thing many :
> For Thou art infinite in one and all."

MICROBES.

The observations of recent years, consequent upon the improvements of the microscope, have resulted in demonstrating the existence and the nature of

myriads of exceedingly minute bodies, of which previously only vague suspicions were entertained.
Notwithstanding their minute size, these microbes
are now known to be important factors in health
and disease, so that, imperfect and incomplete as
our knowledge still remains, medical authorities have
been compelled to admit their existence and influence
in many epidemics.   That objects so minute as to
be invisible to the naked eye should produce such
disastrous results is sufficient to invest them with an
atmosphere of romance, and the names of Bacteria
and Bacilli, which but a few years since were almost
unknown, have become household words.   These
microbes appear, under a high power of the microscope, as small cells of a spherical, oval, or cylindrical
shape, sometimes detached, sometimes united in
pairs, or in chains or chaplets.   So minute are they
that from five hundred to two thousand of them
must be placed end to end in order to attain the
length of a millimetre, which is not more than the
twenty-sixth part of an inch.

It was about the year 1850 that the presence of
minute rods was observed in the blood of animals
which died of splenic fever, but it was not until
1863 that they were suspected of being the cause
of the disease.   At this time Davaine inoculated
healthy animals with the tainted blood, and ascertained that a very minute dose would produce a
fatal attack of the disease, and the rods could be
discovered in enormous quantities in the blood.
This disease is generally inoculated by the bite of
flies, which have absorbed bacteria from diseased

carcases, or by blood-poisoning through some accidental scratch. Pasteur demonstrated by experiment that the disease is really caused by the microbe,
and his experiments have since been confirmed.
Subsequent to this he turned his attention to its
mitigation or cure. Preservative means were those
upon which he chiefly relied, by a process of vaccina-
tion with the virus of anthrax, or splenic fever.
"Pasteur ascertained that when animals are inoculated with a liquid containing bacteria, of which the
virulence has been attenuated by culture, carried as
far as the tenth generation or even further, their lives
are preserved. They take the disease, but generally
in a very mild form, and it is an important result
of this treatment that they are henceforward safe
from a fresh attack of the disease; in a word, they
are *vaccinated* against anthrax."

This is the theory and practice in all disease of
men and animals originating with microbes, which
have, as yet, been the subject of experiment. It is
stated that "up to April, 1882, more than 130,000
sheep and 2000 oxen or cows had been vaccinated;
and since that time the demand for vaccine from
Pasteur's laboratory has reached him from every
quarter."

It is unnecessary to do more than allude to fowl-
cholera, swine-fever, the typhoid fever of horses,
rabies, glanders, etc., as forms of disease having a
similar origin amongst the lower animals, before
passing to those which affect humanity. In former
times miasma was written and spoken about as
noxious emanations in the form of gas, but *now*

malarious gases are regarded as a myth, and intermittent and malarious fevers are attributed to various
microbes. The marsh fever, or malaria, which is so
common in Sicily, and in the campagna of Rome,
has been ascribed to a *Bacillus*, which is abundantly
found in the blood of patients during the period of
attack. "In the strata of air which float above malarious ground in summer, this microbe is so common
that it is found in abundance in the sweat of the
forehead and hands." The disease known as recur-
rent fever, or relapsing typhus, and known in India
as jungle fever, was discovered to be due to microbes
in 1868. The parasite is thread-like, twisted into
numerous spirals, and animated by very lively movements. Some authorities doubted this disease being
the result of microbes, until the experiments of 1880.
A monkey was successfully inoculated with the disease at Bombay, and after five days the spirals were
found in the animal's blood. Yellow fever has not
as yet been exhaustively studied in the countries in
which it prevails, but it is suspected from certain
symptoms and phenomena that further research will
confirm the parasitic nature of the disease. No two
diseases excite more dread than typhoid and typhus
fever, and in the former of these the presence of
special microbes was observed first in 1871, and
since confirmed and more exactly described. The
bacillus has been observed in the spleen, lymphatic
glands, and intestines. It appears, in the form of
short rods with rounded ends, in the glands which
cover the mucous membrane. Many other bacteria
occur in the intestines when the disease approaches

its end, but the bacillus is the only one found in the
blood, and is really characteristic of the disease.
Many are the causes to which cholera has been attributed. Trouessart says that "the essentially epidemic
and contagious progress of this disease clearly indi-
cates the presence of a microbe, of which the chosen
seat is the intestines, whence it passes with the
patient's fæces, and constitutes the contagious element
in places affected by the epidemic." Koch was the
first to describe the microbe which is considered to
be the producing agent of cholera, and he called it
the "comma bacillus," after its form (Fig. 56). He
cultivated it successfully in several media, but for a

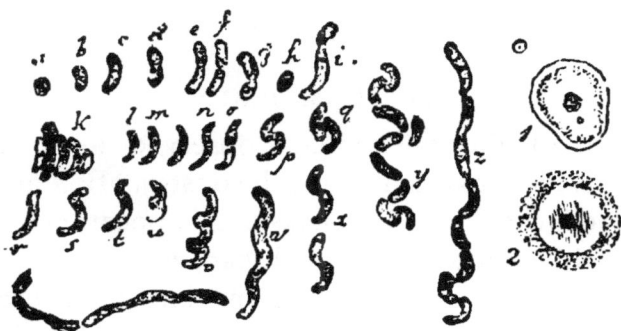

Fig. 56.—Comma Microbe (*Trouessart*).

long while all attempts at inoculating animals by in-
jection of the bacillus failed. Subsequently cholera
was thus produced in guinea-pigs, dogs, etc., which died
at the end of two or three days, with the intestines
containing a number of vigorous comma bacilli. In
1884 Ferran conducted a series of experiments at
Toulon, and professes to have obtained an attenuated
microbe for preventive inoculations. He believes that

U

he succeeded, and, after inoculating himself, he performed the same operation on several of his friends; then on thousands of people in different towns of the province of Barcelona, and throughout Spain.

In like manner microbes have been found in such eruptive diseases as scarlatina, small-pox, measles, etc., whilst extensive details might be added of experiments and researches in diphtheria, croup, whooping cough, influenza, leprosy, tuberculosis, pneumonia, etc., all of which are now believed to be more or less associated with microbes. Enough has been written to show that a large number of human diseases owe their contagious properties to the presence of organisms exceedingly minute, but powerful on account of their numbers and facility of increase.

BLUT IM BRODE.

None of the phenomena observed in lower organisms have excited so much consternation amongst the uneducated as the occasional appearance of red stains, like blood-spots on bread, and other articles of food. "In 1819," we learn that "a peasant of Liguara, near Padua, was terrified by the sight of blood-stains scattered over some polenta which had been made and shut up in a cupboard on the previous evening. Next day similar patches appeared on the bread, meat, and other articles of food in the same cupboard. It was naturally regarded as a miracle and warning from heaven, until the case had been submitted to a Paduan naturalist, who easily ascertained the presence of a microscopic plant, which Ehrenberg had likewise found at Berlin in analogous

circumstances, and which he named *Monas prodigiosa.*
We now term it *Micrococcus prodigiosus.* It has been
observed, not only on bread, but on the Host, on
milk, paste, and on all alimentary or farinaceous
substances exposed to damp heat."[1]

The spherical cells of which the substance is composed, examined under the microscope, are seen to
be filled with a reddish oil, which gives to them a
peach-blossom tint, and when transferred to raw
meat they assume a splendid fuchsia colour, resembling spots of blood. The plant is only developed
in the dark, and the nitrogen necessary for its nutrition must be derived from the air, especially when
developed upon bread. About 1886 an epidemic
appearance on the Conti-
nent was attributed to this
source. Pieces of cooked
meat presented a singular
carmine-red colouration, and
stained vividly the fingers or
linen with which they came
in contact. These phenomena
prevailed regularly for a period

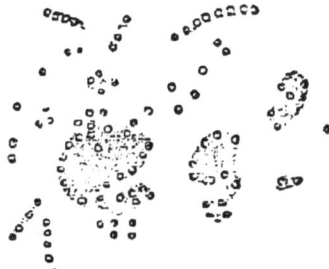

FIG. 57.—*Micrococcus prodigiosus*
(W. B. Grove).

of three months. Food cooked overnight was found
the next morning covered with red patches, and it
then underwent rapid alteration (Fig. 57). Coincident
with a sudden and considerable fall in the temperature, the epidemic ceased, and has not reappeared.[2]
Rev. M. J. Berkeley has also referred to the same

[1] E. L. Trouessart, "Microbes, Ferments, and Moulds," p. 127.
(1889).
[2] *Pharmaceutical Journal* (Jan. 29, 1887), p. 610.

phenomenon, in the following manner. "In the hot days of July, 1853, provisions which were cooked in the evening were in some cases the next morning covered with this production. The only instance of similarly rapid development is that of yeast globules, and it is there probably that we must look for the true solution of the question as to its real nature." And again, he writes, "The rapidity with which it spreads over meat, boiled vegetables, or even decaying agarics, is quite astonishing, making them appear as if spotted with arterial blood ; and what increases the illusion is that there are little detached specks, exactly as if they had been squirted in jets from a small artery. The particles of which the substance is composed have an active molecular motion, but the morphosis of the production has not yet been properly observed, and till that is the case it will be impossible to assign its place rightly in the vegetable world. [This has since been determined.] Its resemblance to the gelatinous specks which occur on mouldy paste or raw meat in an incipient state of decomposition satisfy me that it is not properly an alga."[1] Another observer says, "I observed at table the under surface of a half-round of salt beef, cooked the day before, to be specked with several bright carmine-coloured spots, as if the dish in which the meat was placed had contained minute portions of red-currant jelly. On examination the next day, the spots had spread into patches of a vivid carmine-red stratum of two or more inches in length. With a simple lens the plant appears to consist of a gelatinous sub-

[1] Berkeley, "Introduction to Cryptogamic Botany," p. 114.

stratum of a paler red, bearing an upper layer of a
vivid red hue, having an uneven or papillate surface.
The microscope shows this stratum to consist of
generally globose cells immersed in, or connected by,
mucilaginous or gelatinous matter. The cells vary
in size and contain red endochrome; as far as I can
observe they consist of a single-cell membrane, and
contain a nucleus."[1] Both the foregoing accounts
were written very many years ago, and before the
organism was thoroughly understood.

Fresenius, in his "Beitrage," records the results of
his examination of these spots, which are called by
the Germans "Blut im Brode," from which the following is an extract.

"I took four boiled potatoes, and placed them in
a drawer, having previously rubbed two of them
slightly here and there with the red substance.
After about twenty-four hours the two potatoes
which had not been rubbed, and which had not
been in immediate contact with the other two, were
affected with fresh spots of the red substance, whilst
the spots upon the two which had been rubbed had
increased in extent. The spots showed themselves
in the form of irregular groups of blood-red drops of
different size, which in some places were distinct and
in others had run into one another. The individual
bodies of which the spots consist are mere molecules,
their diameter varying from one two-thousandth to
one four-thousandth of a line. They are mostly
round, occasionally oval, and sometimes slightly constricted in the middle by way of preparation for

[1] H. O. Stephens, in *Ann. and Mag. of Nat. Hist.* (1853), p. 409.

increase by division into two small round cells. By
far the greater number of them, when brought under
the microscope in a drop of water, remain at rest—
they lie close together in large numbers ; when they
are more dispersed in the fluid, they have a motion
which is not distinguishable from ordinary molecular
motion. When the drop of water moves they are
carried mechanically over the stage like other molecules, and when this motion ceases they remain at
one spot in a sort of quivering state until a fresh
current carries them in another direction. If the eye
be kept carefully upon a part of the stage where the
small bodies are thinly dispersed, it will be observed
that they passively follow the current of the water,
nor, when the current has become sluggish, or has
even altogether ceased, are individual bodies ever
seen to detach themselves from the group and take
a contrary direction, which real monads would do
with great activity."

BLOOD RAIN.

It is not always easy to determine from imperfect
description what is the precise organism which causes
the colour and gives rise to the phenomena of " blood-
rain." Ehrenberg discovered and described a minute
organism, which came to be called *Ophidomonas sanguinea*, and, in more recent times, *Spirillum sangui-
neum*, which has the credit of causing blood-rain (Fig.
58). It is one of the *Schizomycetes*, or minute beings
to which the *Bacteria* belong. " Like many other
plants, it readily passes from green to red. No one is
surprised by the green scum which covers reservoirs

in summer, since it is so common, but when this colour changes, often in a single night, and passes from green to red, the unaccustomed tint excites wonder, although it is caused by the same plant which was green the day before. If there is a thunderstorm or waterspout, which draws up the red water from the ponds and reservoirs, and discharges it in the form of rain on the surrounding country, we hear of the phenomenon that it rains

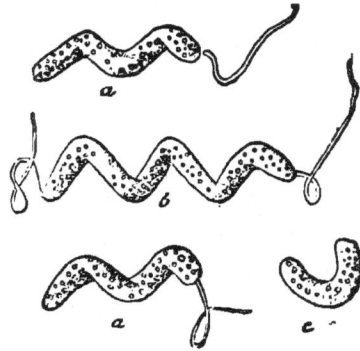

Fig. 58.—*Ophidomonas sanguinea.*

blood, and it would be easy to find in the drops of rain the reddish microbe which imparts this colour to them."[1]

We have in remembrance an incident which occurred in a remote country village some years ago. It was a warm summer day when the report reached us that the villagers were deeply concerned as to a neighbouring horse-pond, which was really almost stagnant, and much reduced by the summer heat. The circumstance which caused alarm was that a large portion of the surface every day assumed a deep blood-red hue, and speculation was rife as to the cause. Out of curiosity we went to see the phenomenon, and saw the irregular, somewhat shifting red patches. At once suspecting the cause, we secured some of the water in a wide-mouthed collecting bottle, and the microscope soon revealed

[1] E. L. Trouessart, "Microbes, Ferments, etc.," p. 128.

myriads of red *Euglena*, which had imparted the colour.

Professor Ehrenberg [1] has given an elaborate historical account of the various reported instances of blood-rain from the most remote times. The first instance adduced dates about fifteen hundred before the Christian Era. It is the plague of blood inflicted upon the Egyptians, as related in the Mosaic history, which prevailed through the whole land of Egypt, continuing three days and three nights. The second occurred about 1181 B.C., the time of Æneas and Dido, as related by Virgil (" Æneid," iv. 454) :—

> "Strange to relate ! for when, before the shrine
> She pours in sacrifice the purple wine,
> The purple wine is turned to putrid blood." [2]

The third was about 950 B.C., as described by Homer :—

> "Even Jove, whose thunder spoke his wrath, distill'd
> Red drops of blood o'er all the fatal field."
>
> *Iliad*, xi. 52.

And again—

> "Then touch'd with grief, the weeping heavens distill'd
> A shower of blood o'er all the fatal field."
>
> *Iliad*, xvi. 459.

The fourth dates about 910 B.C., and is the instance of bloody waters mentioned in connection with the victory over the Moabites (2 Kings iii. 21–23).

He then mentions the rain of blood in the time of

---

[1] C. G. Ehrenberg, " Passat-Staub und Blut-Regen." (Abhander Akad, Berlin, 1847–1849.)

[2] This quotation is hardly so *apropos* as the others which follow.

Romulus, as related by Livy, and after that passes to subsequent periods. In the year 1222 a blood-rain fell at Rome for one day and night. In 1623 there was another blood-rain at Strasburg, which happened on the 12th of August, between the hours of four and five in the afternoon. In 1755, on the 14th of October, at eight o'clock in the morning, a warm sirocco wind was blowing at Locarno, near Lago Maggiore. At ten o'clock the air was filled with a red mist, and at four o'clock p.m. there was a blood-red rain, which left a reddish deposit, equal to one-ninth of its mass. There fell nine inches of this rain in one night. About forty square German leagues were covered with this bloody rain, which also extended on the north side of the Alps into Suabia, and nine feet of reddish snow fell upon the Alps. Supposing that the deposit averaged but two lines in depth, there would be for each square English mile an amount equal to 2700 cubic feet. But actual measurement gave for the depth, in some places, about one inch.

Mention is also made that Humboldt, when in Paramo, on the way from Bogota to Popayan, at a height of 14,700 feet, observed a red hail, which fact was published by him in 1825.[1]

And yet no one can determine what the colouring matter was in each of these instances. Possibly in some of them it was simply the rust-coloured dust, which gives the colour to some of the "dust showers" which have from time to time been recorded. This

[1] Summary of Ehrenberg's paper on "Blood Rain," in *Silliman's Journal*, 1851, p. 372.

view appears to have been held by Ehrenberg, who
includes these latter instances of blood-rain with
his dust showers, in his volume on this subject.   One
fact may, however, be arrived at, and that is the
possibility of water being coloured, even of a blood-
red colour, by the presence of minute organisms
belonging either to the fungi or to the algæ ; that
there is nothing miraculous or extravagant in such
an event ; and although worthy of record as a phenomenon of exceptional occurrence, it is not one
that is open to doubt on a suspicion of impossibility,
or one which should alarm the superstitious as a
portentous omen.   Taken in connection with red
snow, dust showers, and such like occurrences, it will
be conceded that inquiry and investigation is only
necessary to trace them to their sources, and demonstrate not only their possibility, but that they result
from perfectly natural causes.   However much local
colouring may distort facts, and mere report be unsatisfactory, without collateral evidence, there is no
doubt that most of the recorded instances of so-called
blood-rain did take place, although it is hopeless
to attempt to explain them in the absence of scientific
data.   One such mystery cleared up, and denuded
of its fictitious surroundings, is quite sufficient to
predicate for the rest some foundation of fact, and
remove them from the region of romance.

GINGER-BEER PLANT.

A peculiar substance is known in country districts
as the " ginger-beer plant," because it is used to produce home-made ginger-beer.   By examination it is

believed that this substance consists of "low" or sedimentary yeast, mixed with *Saccharomyces mycoderma*, which forms "ropy beer" (Fig. 59), together with various species of *Bacillus*, and, in addition, the

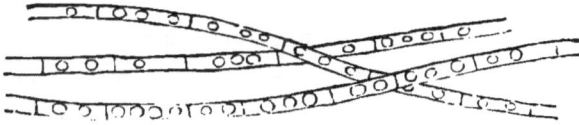

Fig. 59.—*Saccharomyces mycoderma* (W. B. Grove).

"mucor-ferment" of Pasteur (Fig. 60), which is considered by him as a submerged vegetating form of *Mucor racemosus*.[1] The following may be accepted as a fair description of the object in question. In appearance it resembles ordinary tapioca as sold in commerce. "It is white, in many pieces and variable in form, and when handled is found to be fairly firm,

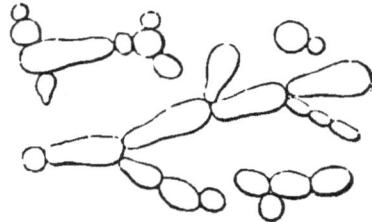

Fig. 60.—Mucor-ferment (W. B. Grove).

somewhat slippery, and very light. Still, its normal position seems to be at the bottom of the bottle which contains it, until, in the process of change which goes on within, the pieces ascend to the surface, and there remain for a time, and presently return to the bottom, so that, whilst the contents are said to be working, these tapioca-like pieces of fungus are constantly ascending and descending. It is evident that carbonic-acid gas is also being fast formed from the sugar, and this too is constantly ascending

[1] W. B. Grove, B.A., "Synopsis of Bacteria and Yeast Fungi," p. 67. (1884.)

to the surface in small globules. Even portions of
the ginger at times are affected with the motions
going on, and ascend and descend, as do the lighter
and more buoyant fungus. This so-called plant
certainly cannot be credited with the production of
ginger-beer. It simply produces an aerated liquid in
this way—that the fungus seems to convert the sugar,
or saccharine, into alcohol and water, accompanied
by the emission of carbonic-acid gas ; and evidently
as the pieces become charged with the gas, they
become buoyant and ascend, falling again when the
buoyancy is lost. In twenty-four hours the sugar
and water are converted into a pleasant sub-acid
beverage, and either ginger or essence of lemon
creates a non-effervescent drink far more pleasant
than is ordinarily manufactured. Fresh water and
sugar should be added about every twenty-four
hours." [1]

Mr. Marshall Ward has recently turned his attention to this subject,[2] and made numerous investigations and cultures, the results differing from those of
previous observers. He finds that the substance
consists essentially of a symbiotic association of a
saccharomycete and schizomycete, mixed with various
other forms. These he groups as follows:—

1. The essential organisms are a yeast, which
turns out to be a new species, allied to *Saccharomyces ellipsoideus* (Reess), which he proposes to
call *Saccharomyces pyriformis ;* and a bacterium, also

---

[1] *Gardener's Chronicle*, June 7, 1884, p. 748.

[2] "The Ginger-Beer Plant, etc.," by H. Marshall Ward, M.A., in
Proceedings of Royal Society, Jan. 20, 1892, p. 261.

new and of a new type, named by him *Bacterium vermiforme.*

2. Two other forms were met with in all the specimens examined—*Mycoderma cerevisiæ* (Desm) and *Bacterium aceti* (Kutz).

3. As foreign intruders, more or less commonly occurring in the various specimens, about fifteen are enumerated, including the common beer-yeast and three or four other unknown yeasts, and the common blue mould, *Penicillium glaucum.*

This curious complex substance, according to each observer, seems to consist essentially of a "yeast-fungus," or saccharomycete, and a schizomycete, mixed with several less important and probably extraneous organisms, the difference being in the names respectively assigned to them.

VINEGAR PLANT.

A peculiar flabby-looking substance has long been known in rural districts, as employed in the domestic manufacture of vinegar, under the name of "Vinegar Plant." "This plant, which has a tough gelatinous consistence, when placed in a mixture of treacle, sugar, and water, gives rise to a sort of fermentation, by which vinegar is produced. After six or eight weeks the original plant can be divided into two layers, each of which acts as an independent plant, and when placed in syrup continues to produce vinegar, and to divide at certain periods of growth. The vinegar thus produced is always more or less of a syrupy nature, and, when evaporated to dryness, a large quantity of saccharine matter is left. Various conjectures have been hazarded as to the origin of the so-called vinegar-plant, some stating that it came from South America or other distant regions, and others that it is a spontaneous production." Lindley, upon the authority of the Rev. M. J. Berkeley, considered it a peculiar sterile form of *Penicillium crustaceum*, or blue mould, which, on being deprived of a superfluity of moisture, sent up fertile threads and produced the fructification of "blue mould." When immersed in fluid, "in place of producing the usual sporiferous stalks, the mycelium increases to an extraordinary extent; its cellular threads interlacing together in a remarkable manner, and producing one expanded cellular mass, with occasionally rounded bodies like spores, in its substance." This is the yeast-like budding characteristic of saccharomycetal plants. " If the plant is allowed to continue growing it forms numerous plates, one above the other." Thus much Professor Balfour narrated forty years ago,[1] to which he added the results of certain experiments with moulds when immersed in syrup with similar results. " Some mould that had grown on an apple was put into syrup in March, and in the course of two months there was a cellular flat expanded mass formed, while the syrup was converted into vinegar. Some of the original mould was seen on the surface in its usual form. Some mould from a pear was treated in a similar way with the same result; also various moulds growing on bread, tea,

---

[1] " On the Growth of Various Kinds of Mould in Syrup," by J. H. Balfour, in *Trans. Bot. Soc. Edin.* (1852), vol. iv. p. 143.

and other vegetable substances; the effect in most cases being to cause a fermentation which resulted in the production of vinegar."

"In another experiment a quantity of raw sugar, treacle, and water were put into a jar, without any plant being introduced, and they were left untouched for four months. When examined, a growth like that of the vinegar-plant was formed. The plant was removed and put into fresh syrup, and again the production of vinegar took place."

It would appear from experiment, that when purified white sugar alone is used to form syrup, the plant when placed in it does not produce vinegar so readily, the length of time required for the changes varying from four to six months.

Mr. H. J. Slack has also published some observations on this singular organism, which may be quoted in illustration of the views which have prevailed at different periods during the past half century. He says,[1] "The 'Micrographic Dictionary' observes that from various observations, the vinegar-plant may be regarded as the mycelium of *Penicillium glaucum* vegetating actively, and increasing also by crops of gonidia or gemmæ. It adds that 'the moniliform growth is at the same time scarcely distinguishable from the yeast-plant by any satisfactory character, and repeated observations strongly impress us with the idea that these objects are all referable to one species—the vinegar-plant being the form of vegetative growth taking place at low or ordinary temperatures in highly saccharine liquids; while the true

---

[1] "The Vinegar Plant," by H. J. Slack, in *Intellectual Observer.*

yeast-plant, or torula, is formed in the more rapid fermentation, taking place at more elevated temperatures.'[1] The vinegar-plant, commonly so called, is a tough, leathery mass, often used by private families to make vinegar out of a solution of sugar and treacle; but the same plant exists in a more delicate and diffused form when other modes of vinegar-making are employed ; and M. Pasteur shows that it coats the shavings or twigs over which some manufacturers cause a suitable liquid to flow when they desire to promote its acetification. If a thin piece of the large tough vinegar-plant is examined microscopically, a moderate power suffices to show what the 'Micrographic Dictionary' describes, namely, an unorganized jelly and cellular structures of many shapes, often resembling coherent cells of yeast, others being like oïdium, etc. It is also, in those I have examined, easy to see something like an entangled mass of minute threads ; but, when this structure is carefully treated, myriads of bacterium bodies appear, and are found to constitute the chief bulk of the plant itself. I am not aware that these bacterium bodies have been described before, and I will therefore explain the process by which I became acquainted with their existence. First, a small thin strip of the vinegar-plant should be torn off from any part on or below the surface. This should be placed upon a glass slide, moistened with a drop of water, and stretched out by means of a camel-hair pencil-holder, and the back of a penknife. When reduced

---

[1] "This theory is doubtful, as the alcoholic fermentation goes on slowly and at low temperatures with the German sediment yeast."

by stretching and squeezing to a film not exceeding
a 250th or a 300th of an inch thick, cover with thin
glass, press steadily, and view with a high power,
using the achromatic condenser and a small stop.
An immense quantity of little thread-like bodies will
then be seen ; and, if the power be sufficient and the
illumination carefully adjusted, a beaded form will
appear sufficiently often to indicate the class of object
to which they belong. That the little bacterium
bodies have not been obtained by tearing big ones
to pieces in the stretching process will be plain upon
examination of the plant in different conditions of
extension and thickness. I have employed in this
investigation Smith & Beck's one-twentieth and
second eye-piece, giving a magnification of 1750
linear. A minute drop of solution of iodine, followed
by another minute drop of dilute sulphuric acid,
facilitates the view of the beaded structure. The
bacterium bodies probably give rise to the tough
mucus in which they are involved, and the unorganized mass in which they are embedded in the
vinegar-plant may only differ in density from the
more delicate material in which the same kind of
bodies are enveloped when they form a pellicle on
the surface of infusions, or adhere in a more or less
globular shape. The yeast-plant is shown to be
intimately connected, if not identical with, several
vegetable forms to which distinct names have been
assigned ; but both upon botanical and chemico-
physiological grounds, it would be very interesting
to ascertain to what extent that form of it known as
the vinegar-plant is associated with bacterium bodies.

X

It is not enough that in one or two cases we find quantities of these bodies present when alcohol or saccharine matter is converted into vinegar; the question is, are they always present, and do they seem to be the particular agents by which the vinegar-making is carried on? When vinegar is obtained from a saccharine solution, the cane sugar is converted into grape sugar, the grape sugar into alcohol, and the alcohol into vinegar. Thus the vinegar-plant appears to perform the double function of first alcoholizing and then acetifying the solution. Do the yeast-like cells accomplish one portion of this task, and the bacterium bodies the other? The mycoderm of wine does not in its ordinary state give rise to vinegar. Its own vital processes merely supply the means by which the changes incidental to vinous fermentation take place, but it occasionally happens that brewers are greatly teased by the acetous fermentation of their beer occurring after the alcoholic change has finished. I have heard of several instances this summer in which great annoyance has been experienced from this cause, and it would seem either that spores of the vinegar-plant were diffused to a greater extent than usual, or that portions of yeast remaining in the beer had developed into the vinegar-plant form. M. Pasteur's view of fermentation does not coincide with the common statement that the yeast-plant merely separates sugar into carbonic acid and alcohol—at any rate he does not represent that as the entire process, because he tells us that when experiments were performed in close vessels containing, besides the fermenting liquid, a known quantity of air, it was found that the vinegar-plant took oxygen from the air, and therewith converted the alcohol into acetic acid ; and that the mycoderm of wine converted the alcohol into water and carbonic acid. Thus both act as oxydizers, and it is well known that if the vinegar-plant be left in fluid after it has transformed the sugar or alcohol into vinegar, it then burns up the vinegar and leaves the housewife or other manufacturer, who had neglected to remove it at the right period, only dirty water for her pains."

SPONTANEOUS GENERATION.

It is not so many years since many estimable men were seduced by the superficial plausibility of the theory then promulgated to give their adherence to the doctrine of spontaneous generation. Down to the seventeenth century it was imagined that certain animal and vegetable forms did, in the process of decomposition, originate or evolve insect life. All this was done in perfect good faith ; it was accepted as a hereditary tradition, without much thought or investigation. For instance, it was held that if a piece of meat was placed in the sun, and allowed to undergo putrefaction, the grubs, which soon appeared, were the result of a power of spontaneous reproduction which the meat possessed. Thus we could find in old books various formulæ for the manufacture of animal or vegetable compounds, which would result in the production of particular forms of animal life. This is all matter. of history and historical fact. Amongst others a distinguished Italian, named Redi,

took up the subject when almost everybody believed
in it or accepted it, including the celebrated Dr.
Harvey. And although this latter is often quoted
as an opponent of spontaneous generation, he undoubtedly believed in it. And yet Dr. Harvey
contended that every living thing came from an egg,
or ovum ; by which he intended to convey, as Professor Huxley has expressed it, that "every living
thing originated in a little rounded particle of organized substance." Probably the enunciation of this
proposition gave rise to the impression that Harvey
was an opponent of the generally accepted view. It
was at this time that Redi made his attack in a very
simple manner. "He merely covered the piece of
meat with some very fine gauze, and then exposed it
to the same conditions. The result of this was that
no grubs or insects were produced ; he proved that
the grubs originated from the insects who came and
deposited their eggs in the meat, and that they were
hatched by the heat of the sun. By this kind of
inquiry he thoroughly upset the doctrine of spontaneous generation, for his time at least."

This occurred at a period when the microscope, as
a scientific instrument, was unknown. But when that
discovery was sufficiently elaborated to show that an
immense number of minute things could be obtained,
almost at will, from decaying animal and vegetable
matter, the theory of spontaneous generation revived
again. If we look into the early guides to the use of
the microscope, we shall find instructions for decoc-
tions of ordinary black pepper, or hay, which if
steeped in water, would in the course of a few days

swarm with an immense number of minute animalcules swimming in all directions. Associated with
this revival of the old theory were the names of
Needham in England, and Buffon in France. They
said that " these things were absolutely begotten in
the water of the decaying substances out of which the
infusion was made. It did not matter whether you
took animal or vegetable matter, you had only to
steep it in water and expose it, and you would soon
have plenty of animalcules. They made an hypothesis about this which was a very fair one. They
said, this matter of the animal world, or of the higher
plants, appears to be dead; but in reality it has a
sort of dim life about it, which, if it is placed under
fair conditions, will cause it to break up into the
forms of these little animalcules, and they will go
through their lives in the same way as the animal or
plant of which they once formed a part."

An Italian naturalist, Spallanzani, then entered
the field in opposition to the theory, as his compatriot
Redi had previously done, and the dispute waxed
hot and strong ; for although Spallanzani did not
make good his views that the process of production
might be stopped by boiling the water, he was believed to be on the right side. Experiments were
being made, first on one side and then on the other,
but in all cases not entirely satisfactory. The
tendency, however, of all was to show that infusoria
were developed from little spores or eggs, which
were constantly floating in the atmosphere, but which
lose their power of germination if subjected to heat.
A period of quiescence then intervened, although

the question was not satisfactorily disposed of; but reappeared with renewed vigour when two eminent Frenchmen, Pouchet and Pasteur, came forward as the champions on either side. This was a long and hotly contested duel, which was watched with intense interest throughout Europe. M. Pasteur verified a number of experiments, criticized and condemned others, and, above all, demonstrated the weaknesses of M. Pouchet's methods, and succeeded in putting his opponent in the wrong.

The position at this point is thus summarized by Professor Huxley[1]: "Not content with explaining the experiments of others, M. Pasteur went to work to satisfy himself completely. He said to himself, 'If my view is right, and if, in point of fact, all these appearances of spontaneous generation are altogether due to the falling of minute germs suspended in the atmosphere,—why, I ought not only to be able to show the germs, but I ought to be able to catch and sow them, and produce the resulting organisms.' He accordingly constructed a very ingenious apparatus to enable him to accomplish the trapping of the 'germ dust' in the air. He fixed in the window of his room a glass tube, in the centre of which he had placed a ball of gun-cotton, which, as we all know, is ordinary cotton-wool, which from having been steeped in strong acid, is converted into a substance of great explosive power. It is also soluble in alcohol and ether. One end of the glass tube was, of course, open to the external air; and, at the other

---

[1] Professor Huxley, "On our Knowledge of the Causes of the Phenomena of Organic Nature," p. 77. (1863).

end of it, he placed an aspirator, a contrivance for causing a current of the external air to pass through the tube. He kept this apparatus going for four and twenty hours, and then removed the *dusted* gun-cotton, and dissolved it in alcohol and ether. He then allowed this to stand for a few hours, and the result was, that a very fine dust was gradually deposited at the bottom of it. That dust, on being transferred to the stage of a microscope, was found to contain an enormous number of starch grains. We know that the materials of our food, and the greater portions of plants are composed of starch, and we are constantly making use of it in a variety of ways, so that there is always a quantity of it suspended in the air. It is these starch grains which form many of those bright specks that we see dancing in a ray of light sometimes. But, besides these, M. Pasteur found also an immense number of other organic substances, such as spores of fungi, which had been floating about in the air, and had got caged in this way.

"He went farther, and said to himself, 'If these really are the things that gave rise to the appearance of spontaneous generation, I ought to be able to take a ball of this *dusted* gun-cotton, and put it into one of my vessels containing that boiled infusion, which has been kept away from the air, and in which no infusoria are at present developed, and then, if I am right, the introduction of this gun-cotton will give rise to organisms.' Accordingly, he took one of these vessels of infusion, which had been kept eighteen months, without the least appearance of life in it, and, by a most ingenious contrivance, he managed to break it open and introduce such a ball of gun cotton, without allowing the infusion or the cotton ball to come into contact with any air but that which had been subjected to a red heat, and in twenty-four hours he had the satisfaction of finding all the indications of what had been hitherto called spontaneous generation.   He had succeeded in catching the germs, and developing organisms in the way he had anticipated.

"It now struck him that the truth of his conclusions might be demonstrated without all the apparatus he had employed.   To do this he took some decaying animal or vegetable substance, such as urine, which is an extremely decomposable substance, or the juice of yeast, or perhaps some other artificial preparation, and filled a vessel, having a long tubular neck, with it.   He then boiled the liquid and bent that long neck into an S shape, or zigzag, leaving it open at the end.   The infusion then gave no trace of any appearance of spontaneous generation, however long it might be left, as all the germs in the air were deposited in the beginning of the bent neck.   He then cut the tube close to the vessel, and allowed the ordinary air to have free and direct access ; and the result of that was the appearance of organisms in it, as soon as the infusion had been allowed to stand long enough to allow of the growth of those it received from the air, which was about forty-eight hours.   The result of M. Pasteur's experiments proved, therefore, in the most conclusive manner, that all the appearances of spontaneous generation arose from nothing more than the deposition of the germs of organisms which were constantly floating in the air. To this conclusion, however, the objection was made, that, if that were the cause, then the air would contain such an enormous number of these germs that it would be a continual fog. But M. Pasteur replied that they are not there in anything like the number we might suppose, and that an exaggerated view has been held on that subject; he showed that the chances of animal or vegetable life appearing in infusions depend entirely on the conditions under which they are exposed. If they are exposed to the ordinary atmosphere around us, why, of course, you may have organisms appearing early. But, on the other hand, if they are exposed to air at a great height, or in some very quiet cellar, you will often not find a single trace of life. So that M. Pasteur arrived at last at the clear and definite result, that all these appearances are like the case of the worms in the piece of meat, which was refuted by Redi, simply germs carried by the air and deposited in the liquids in which they afterwards appear. For my own part, I conceive that, with the particulars of M. Pasteur's experiments before us, we cannot fail to arrive at his conclusions; and that the doctrine of spontaneous generation has received a final *coup de grâce*."

SUMMARY.

ONE important series of facts are to be gleaned from the foregoing, and these establish that the Cryptogamic plants are numerally important items in the great scheme of nature. True, it may be, that a number of them are so minute as not to be appreciable by the naked eye ; but all of them have a history, a birth, a career, and a death. Let the reader infer what conclusions might be drawn and what lessons derived from such facts.

From the estimates, we may fairly assume that the following is not an exaggerated total of the number of cryptogams which are already known to inhabit the surface of the globe, and that the total will be considerably increased when the dark corners of the world are more definitely explored.

Ferns and fern allies	... ...	4600
Mosses and allies ...	... ...	8000
Liverworts	... ...	2000
Lichens	... ...	8000
Algæ of all sections	... ...	10,000
Fungi and pseudo-fungi	... ...	40,000
	Total	72,600

In round numbers we may confidently anticipate that when all the scattered descriptions of Cryptogamic plants have been collected together the total will not fall far short of eighty thousand, of which the fungi, the only section catalogued up to date save the ferns, attain to half of the total number.

Baron von Humboldt, in one of his works, which was first published about three quarters of a century ago, gives an estimate of the number of Cryptogamic plants known at that period, which may be interesting for comparison. "If we estimate," he says, "the whole number of the Cryptogamia hitherto described at 19,000 species, as has been done by Dr. Klotsch, a naturalist possessing a profound acquaintance with the agamic plants, we shall have for the fungi 8000 (of which Agarici constitute the eighth part); for lichens, according to J. von Flotow, of Hirschberg, and Hampe, of Blankenburg, at least 1400; for the algæ 2580; for mosses and liverworts, according to Carl Muller, of Halle, and Dr. Gottsche, of Hamburg, 3800; and for ferns 3250. For this last important result we are indebted to the profound investigations made by Professor Kunze, of Leipzig, on this group of plants. It is a striking fact that the family of the Polypodiaceæ alone includes 2165 of the whole number of described ferns, whilst other forms, as the Lycopodiaceæ and Hymenophyllaceæ, number only 350 and 200. There are, therefore, nearly as many described species among ferns as among grasses."[1] It will be observed that the least amount of increase has occurred in ferns, which have only advanced

[1] Humboldt, "Views of Nature," p. 338.

from 3250 to 4600; but this may be accounted for, not only by their larger size and more imposing appearance, but also from the fact of their having long been a great favourite with collectors and horticulturalists.[1]

It would be strange indeed if nearly eighty thousand different organisms did not present some features of interest to the popular mind, or offer material for thought to those who desire to look beneath the surface of things. The romantic aspect of the cryptogamia is by no means exhausted by the foregoing pages, but perhaps something has been done to show that even amongst the humblest organisms in creation there are marvels and mysteries, lessons and suggestions, facts and phenomena, which are worthy of the attention of young and old, and sometimes almost as fascinating as a fairy tale.

NOTE TO THE ABOVE.—Since the estimate above quoted was printed, Professor Saccardo has issued the following census :—

Filices	...	...	...	...	...	2819
Equisetaceæ, Marsiliaceæ, etc.	...	...	...	565
Musci	...	...	...	...	...	4609
Hepaticæ	...	...	...	...	...	3041
Lichenes	...	...	...	...	...	5600
Fungi	...	...	...	...	...	39,663
Algæ	...	...	...	...	...	12,178

Total 68,475

INDEX.

THE END.

PRINTED BY WILLIAM CLOWES AND SONS, LIMITED, LONDON AND BECCLES

BOOKS BY THE SAME AUTHOR.

FREAKS AND MARVELS OF PLANT LIFE; or, Curiosities of Vegetation. With numerous Illustrations. Post 8vo, cloth boards, 6s.

PONDS AND DITCHES. With numerous Woodcuts. Fcap. 8vo, cloth boards, 2s. 6d.

THE WOODLANDS. With numerous Woodcuts. Fcap. 8vo, cloth boards, 2s. 6d.

TOILERS IN THE SEA; or, Marine Home Builders. With numerous Illustrations. Post 8vo, cloth boards, 5s.

VEGETABLE WASPS AND PLANT WORMS; a Popular History of Entomogenous Fungi. Illustrated. Post 8vo, cloth boards, 5s.

LONDON:

SOCIETY FOR PROMOTING CHRISTIAN KNOWLEDGE,

NORTHUMBERLAND AVENUE, W.C.

Y

PUBLICATIONS

OF THE

Society for Promoting Christian Knowledge.

PUBLICATIONS

OF THE

Society for Promoting Christian Knowledge.

THE ROMANCE OF SCIENCE.

A Series of Books which shows that Science has for the masses as great interest and more edification than the Romances of the day.

Post 8vo, with numerous Woodcuts, Cloth boards.

———◆◆◆———

COAL; and what we get from it.
By Professor R. MELDOLA, F.R.S., F.I.C. 2s. 6d.

COLOUR MEASUREMENT AND MIXTURE.
By CAPTAIN W. DE W. ABNEY, C.B., R.E., F.R.S. 2s. 6d.

DISEASES OF PLANTS.
By Professor MARSHALL WARD. 2s. 6d.

OUR SECRET FRIENDS AND FOES.
By PERCY FARADAY FRANKLAND, Ph.D., B.Sc. (Lond.), F.R.S. 2s. 6d.

SOAP BUBBLES, and the Forces which Mould them.
By C. V. BOYS, A.R.S.M., F.R.S. 2s. 6d.

SPINNING TOPS.
By Professor J. PERRY, M.E., F.R.S. 2s. 6d.

TIME AND TIDE: a Romance of the Moon.
By Sir ROBERT S. BALL. 2s. 6d.

THE MAKING OF FLOWERS.
By the Rev. Professor GEORGE HENSLOW, F.L.S., F.G.S. 2s. 6d.

THE STORY OF A TINDER BOX.
By the late C. MEYMOTT TIDY, M.B.M.S. 2s.

THE BIRTH AND GROWTH OF WORLDS.
By Professor A. H. GREEN, M.A., F.R.S. 1s.

MISCELLANEOUS PUBLICATIONS.

s. d.

ANIMAL CREATION (THE). A popular Introduction to Zoology. By the late THOMAS RYMER JONES, F.R.S. With 488 woodcuts. Post 8vo. ....*Cloth boards* 7 6

BIRDS' NESTS AND EGGS. With 11 coloured plates of Eggs. Square 16mo. ....*Cloth boards* 3 0

BRITISH BIRDS IN THEIR HAUNTS. By the late Rev. C. A. JOHNS, B.A., F.L.S. With 190 engravings by Wolf and Whymper. Post 8vo...........................*Cloth boards* 6 0

EVENINGS AT THE MICROSCOPE; or, Researches among the Minuter Organs and Forms of Animal Life. By the late PHILIP HENRY GOSSE, F.R.S. With 112 woodcuts. Post 8vo. ....*Cloth boards* 4 0

FERN PORTFOLIO (THE). By FRANCIS G. HEATH, Author of "Where to find Ferns," &c. With 15 plates, elaborately drawn life-size, exquisitely coloured from Nature, with descriptive text ....*Cloth boards* 8 0

FISHES, NATURAL HISTORY OF BRITISH: their Structure, Economic Uses, and Capture by Net and Rod. By the late FRANK BUCKLAND. With numerous illustrations. Crown 8vo. ....*Cloth boards* 5 0

FLOWERS OF THE FIELD. By the late Rev. C. A. JOHNS, B.A., F.L.S. New Edition, with an Appendix on Grasses by C. H. JOHNS, M.A. With numerous woodcuts. Post 8vo. ....*Cloth boards* 6 0

FOREST TREES (THE) OF GREAT BRITAIN. By the late Rev. C. A. JOHNS, B.A., F.L.S. With 150 woodcuts. Post 8vo. ....*Cloth boards* 5 0

FREAKS AND MARVELS OF PLANT LIFE; or, Curiosities of Vegetation. By M. C. COOKE, M.A., LL.D. With numerous illustrations. Post 8vo. ....*Cloth boards* 6 0

MAN AND HIS HANDIWORK. By the late Rev. J. G.
WOOD, Author of "Lane and Field," &c. With about
500 illustrations. Large Post 8vo. ....*Cloth boards* 10 6

NATURAL HISTORY OF THE BIBLE (THE). By the
Rev. Canon TRISTRAM, Author of "The Land of Israel."
With numerous illustrations. Crown 8vo....... *Cloth boards* 5 0

NATURE AND HER SERVANTS ; or, Sketches of the
Animal Kingdom. By the Rev. THEODORE WOOD. With
numerous woodcuts. Large Post 8vo. ....*Cloth boards* 5 0

OCEAN (THE). By the late PHILIP HENRY GOSSE, F.R.S.,
Author of "Evenings at the Microscope." With 51
illustrations and woodcuts. Post 8vo.............*Cloth boards* 3 0

OUR INSECT ALLIES. By the Rev. THEODORE WOOD.
With numerous illustrations. Fcap. 8vo. ......*Cloth boards* 2 6

OUR INSECT ENEMIES. By the Rev. THEODORE WOOD.
With numerous illustrations. Fcap. 8vo. ......*Cloth boards* 2 6

OUR ISLAND CONTINENT. A Naturalist's Holiday in
Australia. By J. E. TAYLOR, F.L.S., F.G.S. With Map.
Fcap. 8vo..*Cloth boards* 2 6

OUR NATIVE SONGSTERS. By ANNE PRATT, Author
of "Wild Flowers." With 72 coloured plates. 16mo.
Cloth boards 6 0

SELBORNE (THE NATURAL HISTORY OF). By the
Rev. GILBERT WHITE. With Frontispiece, Map, and 50
woodcuts. Post 8vo...............................*Cloth boards* 2 6

WAYSIDE SKETCHES. By F. EDWARD HULME, F.L.S.,
F.S.A. With numerous illustrations. Crown 8vo.
Cloth boards 5 0

WHERE TO FIND FERNS. By FRANCIS G. HEATH,
Author of "The Fern Portfolio," &c. With numerous
illustrations. Fcap. 8vo. ....*Cloth boards* 1 6

WILD FLOWERS. By ANNE PRATT, Author of "Our
Native Songsters," &c. With 192 coloured plates. In
two volumes. 16mo. ....*Cloth boards* 8 0

MANUALS OF ELEMENTARY SCIENCE.

A Set of Elementary Manuals on the principal Branches of Science
Fcap. 8vo., limp cloth, 1s. each.

ELECTRICITY. By the late Professor FLEEMING JENKIN.

PHYSIOLOGY. By F. LE GROS CLARK.

GEOLOGY. By the Rev. T. G. BONNEY, M.A., F.G.S.

CHEMISTRY. By A. J. BERNAYS, Ph.D., F.C.S.

CRYSTALLOGRAPHY. By HENRY PALIN GURNEY, M.A.

ASTRONOMY. By W. H. M. CHRISTIE, M.A., F.R.S.

BOTANY. By Professor BENTLEY.

ZOOLOGY. By ALFRE NEWTON, M.A., F.R.S.

MATTER AND MOTION By the late J. CLERK MAXWELL

SPECTROSCOPE AN ITS WORK (THE). the late RICHARD A. PROCTO

NATURAL HISTORY RAMBLES.

Fcap. 8vo., with numerous Woodcuts, Cloth boards, 2s. 6d. each.

IN SEARCH OF MINERALS. By the late D. T. ANSTE M.A., F.R.S.

LAKES AND RIVERS. By C. O. GROOM NAPIER, F.G.! Author of "The Food, Use, and the Beauty of British Birds."

LANE AND FIELD. By the late Rev. J. G. WOOD, M.! Author of "Man and his Handiwork, &c.

MOUNTAIN AND MOOR. By J. E. TAYLOR, F.L.S., F.G.! Editor of "Science-Gossip."

PONDS AND DITCHES. By M. C. COOKE, M.A., LL.D.

THE SEA-SHORE. By Professor P. MARTIN DUNCAN, M. (London), F.R.S., Honorary Fellow of King's College, Londo:

THE WOODLANDS. By M. C. COOKE, M.A., LL.D., Auth of "Freaks and Marvels of Plant Life," &c.

UNDERGROUND. By J. E. TAYLOR, F.L.S., F.G.S.

LONDON: NORTHUMBERLAND AVENUE, W.C.;
43, QUEEN VICTORIA STREET, E.C.
BRIGHTON: 135, NORTH STREET.

www.ingramcontent.com/pod-product-compliance
Lightning Source LLC
Chambersburg PA
CBHW021458210326
41599CB00012B/1050